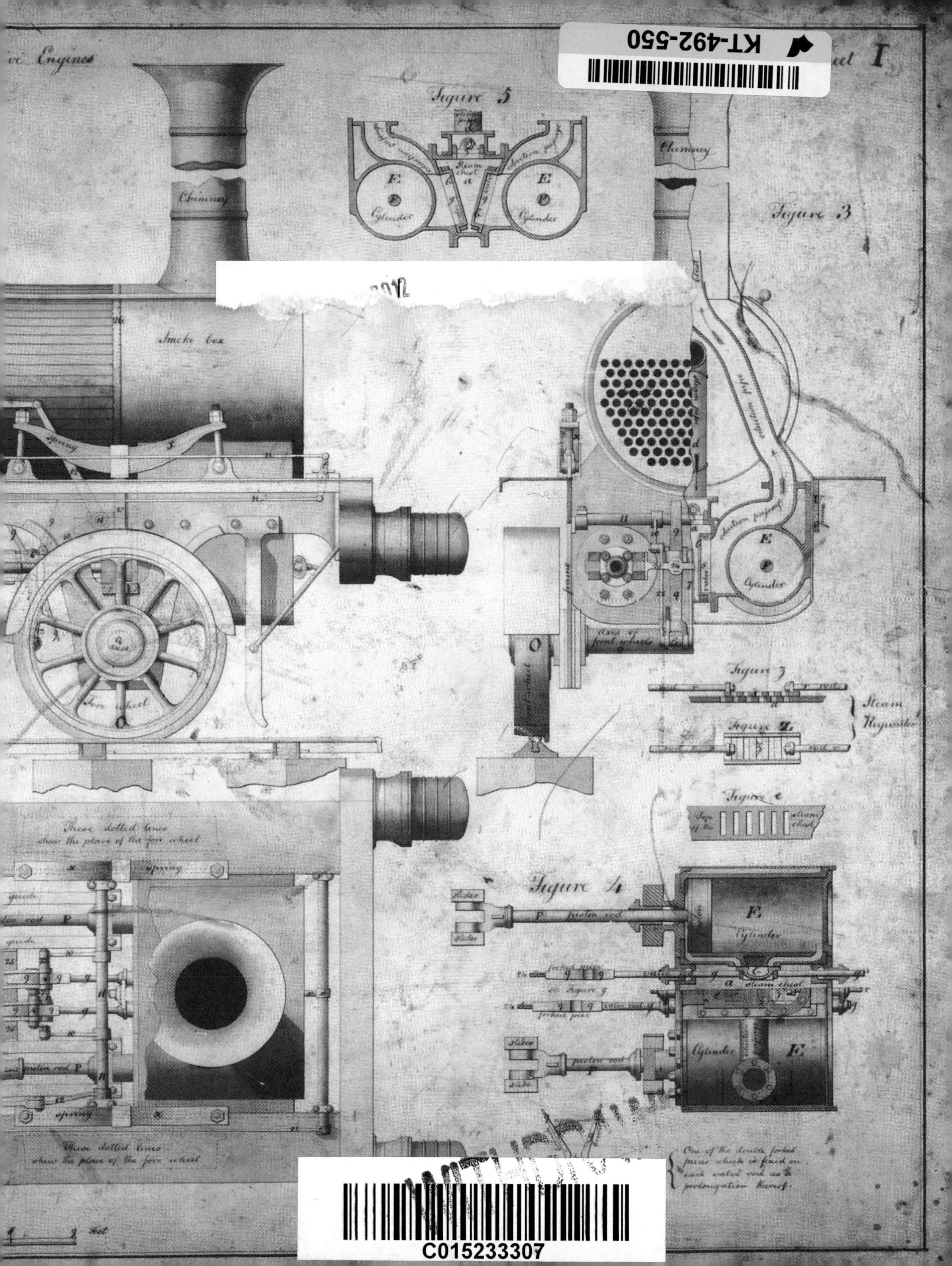

ive Engines
Sheet I
Figure 5
Steam chest
a
E
Cylinder
eduction passage
Chimney
Figure 3
Smoke box
spring
Fore wheel
eduction pipe
steam pipe
frame
axis of front wheels
Steam Regulator
Figure 4
piston rod
Cylinder
steam chest
forked piece
valve rod
Slides
These dotted lines shew the place of the fore wheel
One of the double forked pieces which is fixed on each valve rod as a prolongation thereof.
Feet
KT-492-550
C015233307

GREAT
VICTORIAN
RAILWAY
JOURNEYS

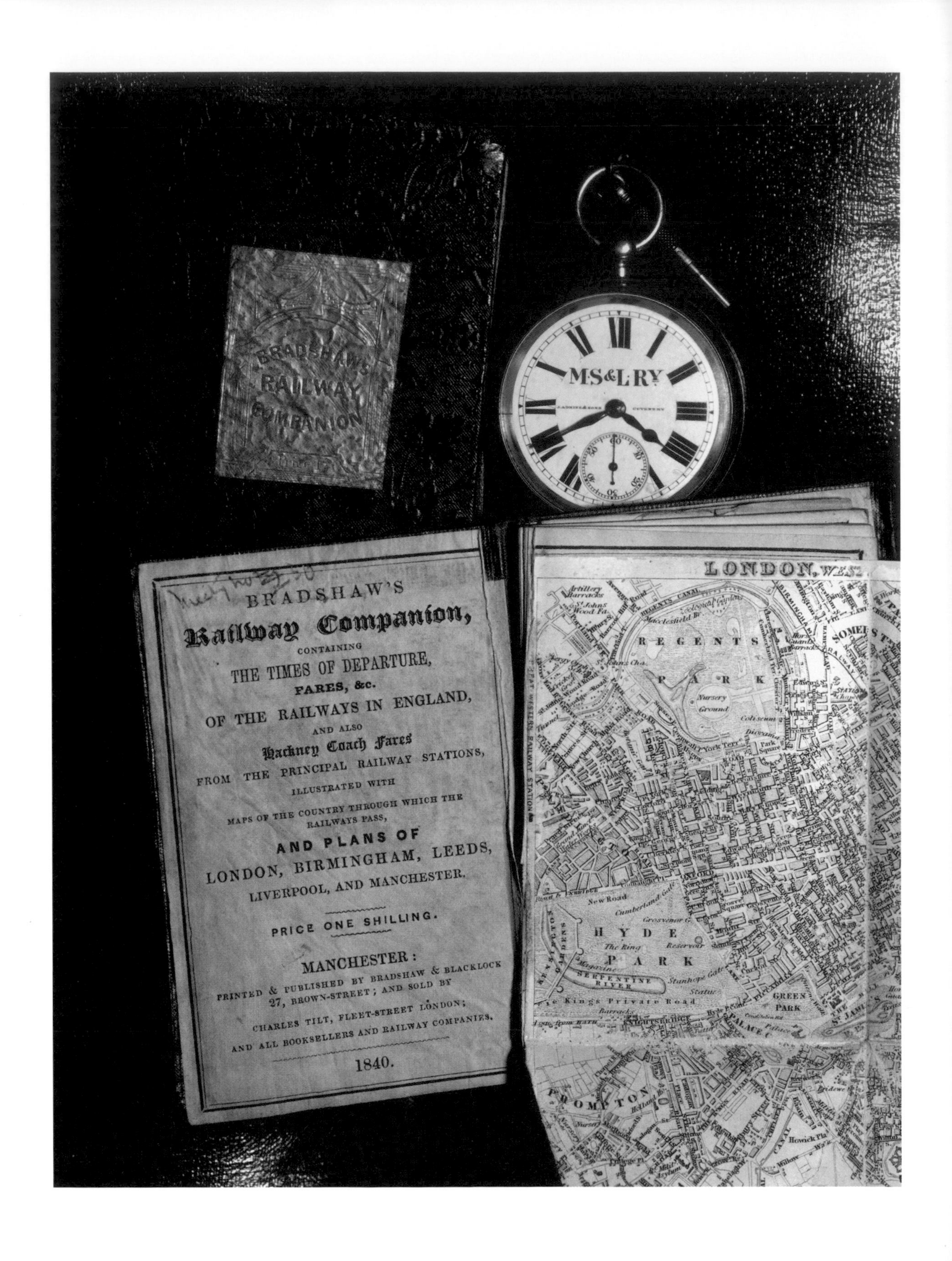
BRADSHAW'S
RAILWAY
COMPANION
M.S.& L.RY
BRADSHAW'S
Railway Companion,
CONTAINING
THE TIMES OF DEPARTURE,
FARES, &c.
OF THE RAILWAYS IN ENGLAND,
AND ALSO
Hackney Coach Fares
FROM THE PRINCIPAL RAILWAY STATIONS,
ILLUSTRATED WITH
MAPS OF THE COUNTRY THROUGH WHICH THE
RAILWAYS PASS,
AND PLANS OF
LONDON, BIRMINGHAM, LEEDS,
LIVERPOOL, AND MANCHESTER.
PRICE ONE SHILLING.
MANCHESTER:
PRINTED & PUBLISHED BY BRADSHAW & BLACKLOCK
27, BROWN-STREET; AND SOLD BY
CHARLES TILT, FLEET-STREET LONDON;
AND ALL BOOKSELLERS AND RAILWAY COMPANIES.
1840.
LONDON, WEST
REGENTS
PARK
Nursery Ground
HYDE
PARK
SERPENTINE RIVER
GREEN PARK
BROMPTON

GREAT VICTORIAN RAILWAY JOURNEYS

KAREN FARRINGTON

Collins

First published in 2012 by Collins

HarperCollins*Publishers*
77–85 Fulham Palace Road
London W6 8JB

www.harpercollins.co.uk

10 9 8 7 6 5 4 3 2 1

Picture captions for front matter: p 2: The first edition of *Bradshaw's Railway Companion* of 1840, with a pocket watch made c. 1870 by J. Atkins & Son of Coventry for a guard on the Manchester, Sheffield and Lincolnshire Railway; pp 6–7 Bradshaw's map of Dublin. Endpapers: Stephenson's 'Long Boiler' Locomotive, 1841: engineering plans, elevations and sections drawn by J. Farley of Guilford Street, London and signed by Robert Stephenson.

A catalogue record for this book is available from the British Library.

ISBN: 978-0-00-745706-9

Printed and bound in Italy by L.E.G.O. SpA

MIX
Paper from responsible sources
FSC® C007454

FSC is a non profit international organisation established to promote the responsible management of the world's forests. Products carrying the FSC label are independently certified to assure consumers that they come from forests that are managed to meet the social, economic and ecological needs of present and future generations.

Find out more about HarperCollins and the environment at
www.harpercollins.co.uk/green

ACKNOWLEDGEMENTS

To write a book about Victorian railways is to chart the changes that transformed Britain from rural idyll to industrial powerhouse. It required extensive reading of published and internet resources. Among the most inspirational books that I used were *A Historical Dictionary of Railways in the British Isles* by David Wragg, *The Age of Steam* by Thomas Crump, *Along Country Lines* by Paul Atterbury, *Ironing the Land* by Kevin O'Connor, *On the Slow Train* by Michael Williams, *Eleven Minutes Late* by Matthew Engel, *Bradshaw's Railway Map 1907*, re-published by Old House Books and *A Book of Railway Journeys*, an anthology compiled by Ludovic Kennedy. On the internet Grace's Guide was a beacon, as were all the superb websites hosted by heritage lines too numerous to mention.

First and foremost, my thanks to everyone at talkbackTHAMES who has supported this book from start to finish. Without the help and advice of John Comerford and Jay Taylor this book would not have been possible. Lucy Butler generously shared her knowledge with me, helping greatly by supplying photographs and interpreting research material. Jacqueline Moreton, Esther Johnson and Cat Ledger provided invaluable back-up. Pam Cavannagh at the BBC once again kept us all on track. And special thanks to Michael Portillo for his support.

Biggest thanks, however, go to Nick Constable for ceaseless support and encouragement. Thanks also to Lewis, Conor and Annie for their consideration.

Karen Farrington
November 2011

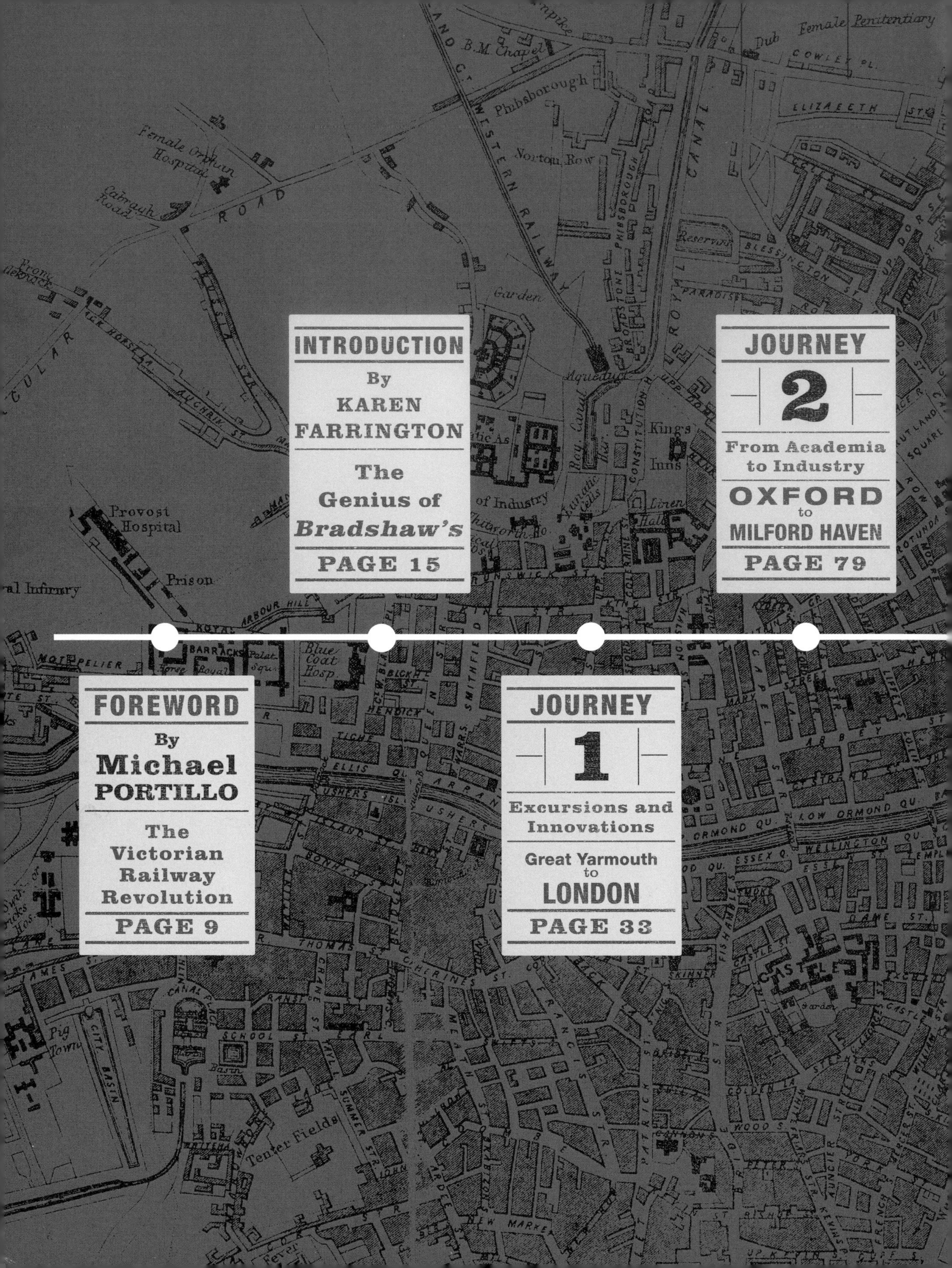

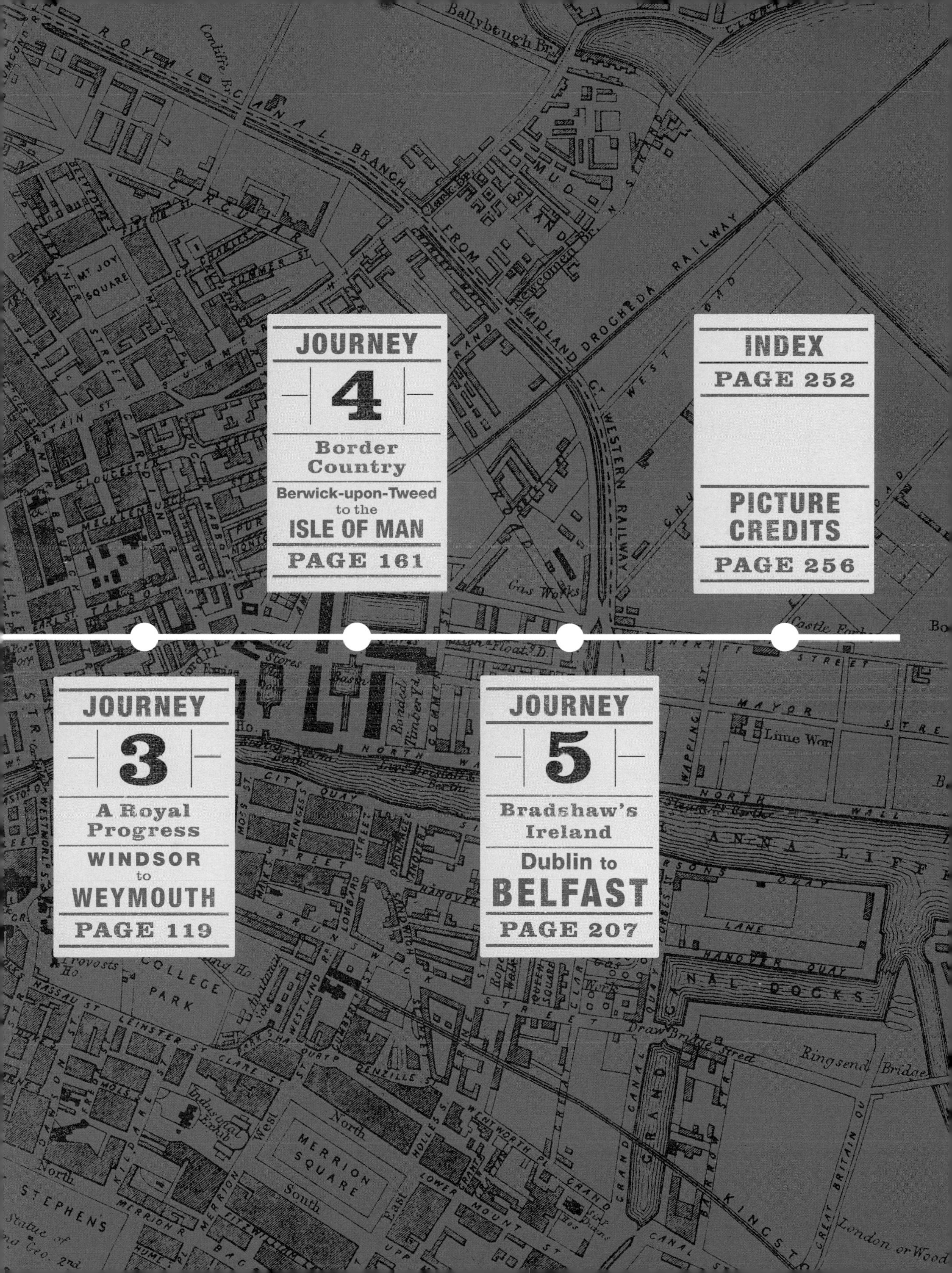

Ballybough Br.
ROYAL CANAL BRANCH
CIRCULAR
MT JOY SQUARE
MIDLAND GT WESTERN RAILWAY
DROGHEDA RAILWAY
Gas Works
CITY QUAY
ANNA LIFFEY
HANOVER QUAY
CANAL DOCKS
COLLEGE PARK
MERRION SQUARE
Ringsend Bridge
GRAND CANAL

THE SOLENT
Gurnet Bay
Thorney Bay
Newtown River
Elmsworth
East Hampstead
Lower Hampstead
YARMOUTH
Newtown
Ningwood Common
Thorley
Thorley St.
Shalfleet
Upper Watchingwell
Lt. Watchingwell
Forest Side
Northwood
Stigwell
PARKHURST FOREST
Lt. Cockleton
Freshwater Ho.
Freshwater
Red Lion
Dog Kennel
Ningwood Green
Swainstone
WEST
Afton House
Afton Farm
Tapnell Farm
Calbourn
Churchills
Castle
Bowcombe
Idlecombe
Westover
Lynch
Rowridge
Chessells
Afton Down
Freshwater Bay
Compton Bay
Mottestone Down
Brixton Down
Limerston Down
Gatcomb
Brook Ledge
Brook Chine
Mottestone
MEDINA
Drapers
Chilton Green
Grange
Brighstone
Brixton
Shorewell
Bull Rock
Chilton Chine
Grange Chine
Barnes Chine
Smallbrook
Billingham
Sutton Chine
Kingston
Brixton Bay
Cowleaze Chine
Atherfield
Stroud Gr.
Atherfield Rocks
Whale Chine
Chale
Walpen Chine
Blackgang Chine
Chale Bay
Broken End

FOREWORD

By Michael PORTILLO

The Victorian Railway Revolution

ORENSTEIN & KOPPEL
No12662 BERLIN-DREWITZ

Since 2009, when I started making televised train journeys using *Bradshaw's Handbook*, I have tried to imagine what impact and impression the arrival of the railways made on British society. Of course every generation since has experienced technological change, which has, if anything, grown faster. As we look back over twenty years, most of us can no longer understand how we lived without mobile phones and the internet. Similarly, a Victorian by the 1850s must have struggled to recall the days when he or she had had to travel by coach, never able to exceed the speed of a horse. There had been a revolution in communications, perhaps similar in scope to the one that we are experiencing today.

But on top of that, Britain underwent a physical metamorphosis. Where before, country lanes meandered, now embankments, tunnels and viaducts scythed through the landscape. Stations sprouted up, resembling Italian palaces, French châteaux or even Gothic cathedrals. Fearsome locomotives belching fire and steam frightened the horses as they sliced through city centres, or engulfed fields in flame as stray sparks set light to hay or stubble.

I remember being in awe and fear of steam locomotives. They were so big and noisy. Their wheels towered above me (especially when I was a boy, of course) and they were given to highly unpredictable behaviour, suddenly releasing clouds of water vapour with a hiss that made me jump. How did the Victorian middle classes in their finery, how did genteel ladies, cope with their sudden proximity to these roaring fireboxes, with being brought face to face with the industrial revolution?

How could they adapt to speed? At the Rainhill trials to choose a source of traction for the Manchester to Liverpool railway, Stephenson's *Rocket* reached 30 miles per hour, a velocity previously unseen and hard even to imagine. The potency of the technology took some getting used to. At the inauguration of that line in 1830, William Huskisson MP, former President of the Board of Trade, left his carriage to greet the Prime Minister, the Duke of Wellington. The *Rocket*, running on the parallel track, struck him as he tried to scramble out of its path. A train transported him to the vicarage at Eccles where he died of his injuries.

When Isambard Kingdom Brunel opened his epic Box tunnel on the Great Western Railway, Dionysius Lardner, a professor of astronomy, was on hand to warn that if the brakes failed the train

PREVIOUS PAGE: **Bradshaw's map of the Isle of Wight.**

On an engine from the Downpatrick Heritage Steam Railway.

would accelerate and the passengers suffocate; and some at least believed him and alighted before the tunnel to continue their journey by more traditional means.

At the beginning of the nineteenth century William Blake wrote of 'dark Satanic Mills'. The railways greatly accelerated the industrial revolution. George Bradshaw was enthusiastic about the railways, and the *Handbook* describes with admiration the dimensions of stations, tunnels and bridges. But the book's depictions of iron-smelting furnaces lighting up the night sky sometimes bring the fires of Hell to mind. Bradshaw's awareness of the social consequences of the industrial revolution suggests that Victorians saw that progress had brought both paradise and inferno.

Of Merthyr Tydfil the *Handbook* says:

> **Visitors should see the furnaces by night, when the red glare of the flames produces an uncommonly striking effect. Indeed, the town is best visited at that time, for by day it will be found dirty... Cholera and fever are of course at home here ... We do hope that proper measures will be taken by those who draw enormous wealth from these [iron and coal] works to improve the condition of the people.**

Quakers like Bradshaw tended to believe strongly that the industrialist owed a duty of care to his workers and their families.

To judge from the Great Western works at Swindon, railway magnates were appropriately paternalistic, housing the employees in model dwellings and supplying free rail travel for a family summer holiday in Devon or Cornwall. Many thousands would depart on charter trains, turning Swindon for a week into a virtual ghost town.

Train travel vastly broadened the horizons of working-class people. The mid-nineteenth century saw an enormous growth in travel and in resorts. As the railway navvies levelled the land, the railways levelled society. Those of modest means, who before train travel might scarcely have travelled beyond a 15-mile radius of home, could leave the smoke and grime of city life and enjoy the 'salubrious' (to use a favourite Bradshaw word) sea or mountain air. Town dwellers could for the first time enjoy fresh sea fish and a range of perishable farm products.

The railways threw up some interesting questions of etiquette, and drove social change. The usual requirement that ladies be chaperoned was difficult to enforce on trains. The new freedoms offered by rail travel contributed to increasing equality between the sexes. Nonetheless, women sometimes journeyed in fear, especially in compartments on trains that had no corridor. Some ladies secreted pins in their mouths as a way to deal with male passengers who in the darkness of a tunnel might attempt to steal a kiss!

Any worries that travelling by train was not ladylike were dismissed when Queen Victoria herself took to the tracks in 1842. Although she was nervous of travelling at high speed, she became a frequent railway passenger. The train conveyed her to Windsor, Osborne House and Balmoral. In 1882 it carried her to Nice, on the first of her nine visits to the Riviera. As though to emphasise how thoroughly the railway had entered national life, after her death on the Isle of Wight a train brought the Queen's coffin home to Windsor.

Her lifetime covers most of the railway building that was done in this country. Poignantly, the last link of the West Highland line between Fort William and Mallaig (since made famous by the Harry Potter films) opened in 1901, the year of the Queen's death. So, if you take a train today, it is extremely likely that you will travel on a route laid down in her day.

British national self-confidence reached its zenith in the Victorian era. The empire was vast and, to Bradshaw's mind, the greatest that the world had ever seen. London was the biggest city on earth and Manchester could take cotton from India, turn it into manufactured garments and sell them back in India below the cost of local products. No hint of colonial guilt affected our patriotic pride.

That Britain was the first country to open an inter-city railway, that geniuses like Brunel and the Stephensons achieved engineering wonders, were evidence of the superiority of British ingenuity, science and entrepreneurship.

As today I travel along Victorian track beds I try to imagine the awe and pride that railways then inspired. Actually, if I look out the carriage window at the stations and viaducts that Victorian engineers erected, imagining it is very easy.

Michael Portillo
2012

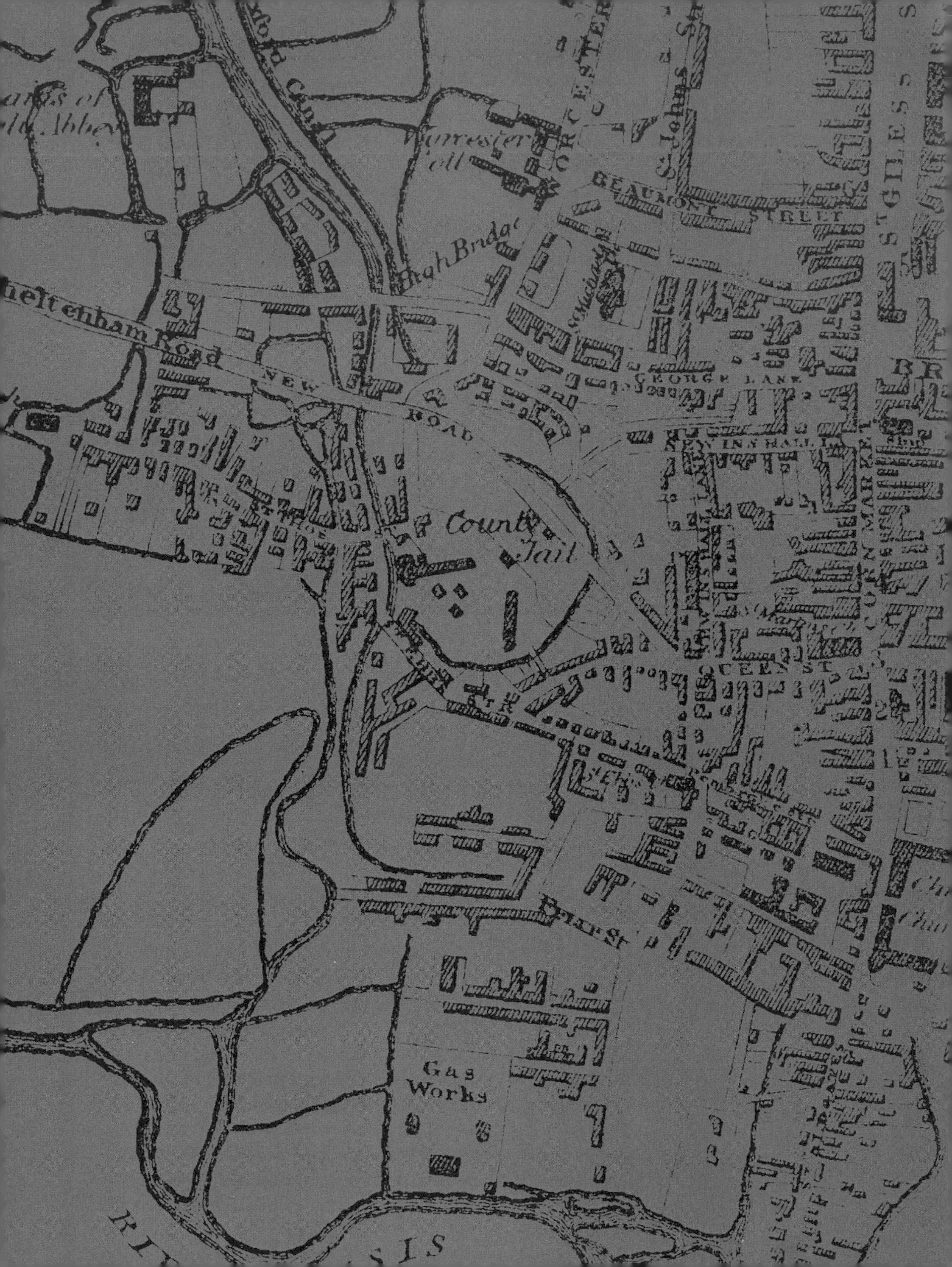

Cheltenham Road
New Road
High Bridge
Worcester
Beaumont Street
St Johns St
St Giles St
George Lane
New Inn Hall L
County Jail
Queen St
Friar St
Gas Works

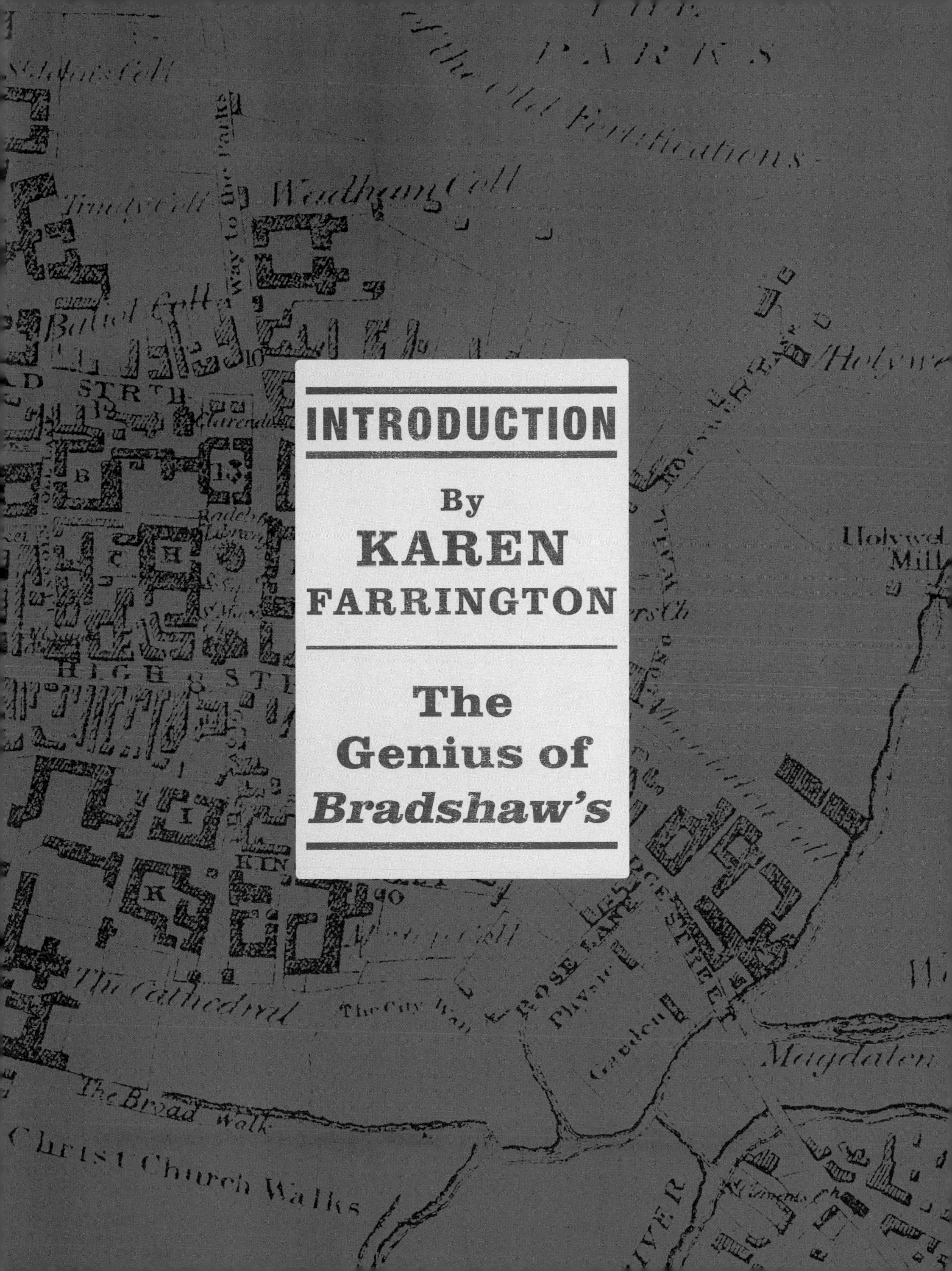

INTRODUCTION

By
KAREN FARRINGTON

The Genius of *Bradshaw's*

When Queen Victoria came to the throne trains had already been running for a dozen years, with modest services along variously sized tracks that amounted to less than 300 miles in length. By then farmers no longer believed trains would set fire to crops or frighten horses to madness, although the sight of a steam engine, rare and awesome, was still sufficiently thrilling to move waving children to the track side.

However, in 1837 a tipping point was fast approaching. During Victoria's reign there was a headlong rush towards railway building that careered out of control, with profit rather than common sense at its heart.

With the rampant growth of railways, regimented tracks treated towns and fields with the same disdain as they splintered into an arbitrary network that reached from city to coast. The UK map was scored with lines that linked one town to the next – or sometimes bypassed entire populations for the sake of a country halt, depending

PREVIOUS PAGE: **Bradshaw's map of Oxford.**

Restored Caledonian Railway 0-6-0 steam engine 828 steams along the Strathspey Railway track between Broomhill and Boat of Garten.

on local politics. Before Victoria's death in 1901 some 18,670 miles of track were on the ground.

One man had the vision to pull the hotchpotch of strands together to make sense of Britain's railways for the travelling public. George Bradshaw was an engraver in Manchester whose first major work was a map of British canals, rivers and railways, published in 1830. In 1838 he began producing railway timetables which, four years later, began appearing as a monthly guide.

One of the first challenges Bradshaw faced was the time lag between, say, London and the West Country, which amounted to some 10 minutes, with separate communities setting their watches by their own calculations of sunrise and sunset. Engineer Isambard Kingdom Brunel had already identified the problem and insisted on standardising clock faces across the Great Western Railway network, using 'railway time'. Bradshaw picked up the baton and made 'railway time' uniform for the whole of Britain for the purposes of his timetabling.

Eventually Bradshaw's publishing stable included a *Continental Railway Guide* and a *Railway Manual, Shareholders' Guide*. But it was the monthly guides that were best sellers and, after Bradshaw's untimely death in 1853 from cholera, the publication continued to appear under his name. Indeed, it was published until 1961.

George Bradshaw, c. 1850s.

The station clock at Crowcombe Heathfield on the West Somerset railway line.

This book focuses on another Bradshaw publication, *Bradshaw's Tourist Handbook*, published in 1866. Inside its pages lies a lyrical image of Britain as it was 150 years ago. There are both hearty recommendations and dire warnings for the Victorian explorer embarking on long-distance domestic journeys. For example, *Bradshaw's* gives an emphatic thumbs-up to Gravesend, with a description that would make any casual reader want to hop on a train and visit:

> **Gravesend is one of the most pleasantly situated and easily attained of all the places thronged upon the margin of the Thames. It is, moreover, a capital starting point for a series of excursions through the finest parts of Kent and has, besides, in its own immediate neighbourhood some tempting allurements to the summer excursionist in the way of attractive scenery and venerable buildings.**

But Cornwall, which today might be considered the brighter prospect of the two for travellers, is altogether less impressive through nineteenth-century eyes:

> **Cornwall, from its soil, appearance and climate, is one of the least inviting of the English counties. A ridge of bare and rugged hills intermixed with bleak moors runs through the midst of its whole length and exhibits the appearance of a dreary waste.**

As intriguing as these descriptive passages are the accompanying advertisements, for 'Davis's Patent Excelsior Knife Cleaner' and 'Morison's Pills – a cure for all curable diseases'. The 1866 fares for day trips and holidays are also advertised.

Excursions were a joyful hallmark of the Victorian age. Many people had more leisure time and spare income than ever before and they broadened their horizons by travelling across the country or the Channel to new resorts and to attend great occasions like exhibitions and major sporting events. For the railway companies, dedicated trains to resorts or events were a profitable way to use idle stock at off-peak times.

A Victorian seaside excursion, c. 1890.

To travel to Dublin via Liverpool from Hull in 1866 cost the first-class passenger 43 shillings. It was 10 shillings less in second class. Return tickets to the Continent via Calais from London were £2 first class and £1 10 shillings in second. Meanwhile, a first-class ticket from Nottingham to Windermere was priced at 42 shillings, and 10 shillings less again for those in second class.

The virtues of Bradshaw's guides in all their forms have long been recognised. Journalist and travel writer Charles Larcom Graves (1856–1944) penned the following verse, reflecting his own cosy association.

When books are pow'rless to beguile
And papers only stir my bile,
For solace and relief I flee
To Bradshaw or the ABC
And find the best of recreations
In studying the names of stations.

Certainly in those days before the advent of through tickets, *Bradshaw's Monthly Railway Guide* was an often-consulted bible for travellers who industriously sought arrival and departure times during detailed journey planning.

No doubt the tripper clutching a Bradshaw guide would feel a rush of nerves as well as excitement in boarding the train, for the notion that railways were dangerous took a long while to subside and railway safety standards were poor by comparison with today. To compound matters, railway companies were reluctant to spend money on even basic safety measures, preferring to reward their shareholders rather than to protect their passengers and staff. Accordingly, brightly painted engines were kept whistle clean but might not be fitted with an adequate number of brakes. The prospect of a rail crash in Victorian times and death by technology invariably filled travellers with fear.

It wasn't until 1875 that a Royal Commission on Railway Accidents was established. It fostered trials on different braking systems, with American-designed systems coming out on top. However, not all railway casualties were linked to crashes, according to railway writer and Royal Commission member William Mitchell Acworth.

> *Passengers tried to jump on and off trains moving at full speed with absolute recklessness. Again and again it is recorded, 'injured, jumped out after his hat', 'fell off, riding on the side of a wagon', 'skull broken, riding on the top of the carriage, came into collision with a bridge', 'guard's head struck against a bridge, attempting to remove a passenger who had improperly seated himself outside', 'fell out of a third class carriage while pushing and jostling with a friend'.*

'SERIOUS ACCIDENTS ... HAPPENED TO PERSONS WHO JUMPED OFF WHEN THE CARRIAGES WERE GOING AT SPEED, GENERALLY AFTER THEIR HATS'

'Of all the serious accidents reported to the Board of Trade,' writes one authority, 'twenty-two happened to persons who jumped off when the carriages were going at speed, generally after their hats, and five persons were run over when lying either drunk or asleep upon the line.'

Putting aside the dread of disaster, there was a lot of general discomfort to overcome, even for those travelling in first class. With as few as four wheels per carriage in the early days of rail, there was no sense of a rhythmic lull, more a violent vibration. Moreover, for decades carriages were unheated, although foot-warmers were available at a price on most routes. Initially only first-class travellers were permitted their use.

Until the mid-1840s third-class passengers were transported in open wagons that were sometimes attached to goods trains. In winter passengers were typically underdressed while subject to wind and rain,

'A Train of the First Class (top) and a Train of the Second Class (bottom)' from *Coloured View of the Liverpool & Manchester Railway*, engraved by S. G. Hughes and published by Ackermann & Co., London in 1832/33.

The Staplehurst Rail Disaster

Author Charles Dickens showed remarkable presence of mind after he was involved in the Staplehurst disaster, but later he wrote about the long-term after-effects he suffered.

Charles Dickens (1812–1870) in middle age.

On 9 June 1865 Dickens was returning from France aboard the boat train, heading for Charing Cross. Along the way, at Staplehurst in Kent, track was being renewed on a railway bridge by a group of 'gangers' or 'waymen', employed by the thousand across the country to repair and maintain track. (The remains of line-side huts where equipment was stored and tea was brewed for this unsung army are sometimes still visible today.)

The train was known as 'the tidal', because its timing varied with the tide that brought the ferries to their berth. According to foreman John Benge's calculations, it would roll into view at 5.20 p.m. In fact, the train was two hours earlier than that and was bearing down on the fractured bridge at a speed of 50 mph. It was impossible to return the necessary track to its position in time.

With rails missing, the locomotive ploughed into the mud lying 10 feet below. The first coach almost went after it but lurched to a halt, balanced precariously between engine and bridge. Its coupling with the rest of the train snapped and another five coaches, travelling with momentum, piled into the gap, landing at all angles.

Ten passengers died and 49 people were injured. Dickens, who was in the first carriage, was uninjured and climbed out of the window. The author tended the wounded as best he could. One man died in his arms and he saw several other dead bodies as he picked his way cautiously through the injured, offering water and brandy.

He wrote about the shocking experience in a letter:

> *No imagination can conceive the ruin, or the extraordinary weights under which people were lying, or the complications into which they were twisted among iron and wood, and mud and water.*
>
> *I have a – I don't know what to call it – constitutional presence of mind and was*

The Staplehurst Rail Disaster: Dickens, Ellen Ternan and her mother escaped unhurt, but he suffered psychologically to the end of his life.

not in the least flustered at the time ... But, in writing these scanty words of recollection, I feel the shakes and am obliged to stop.

The crash left him at a low ebb. Three years later he wrote:

To this hour I have sudden rushes of terror, even when riding in a Hansom cab, which are perfectly unreasonable but quite insurmountable ... my reading secretary and companion knows so well when one of these momentary seizures comes upon me in a railway carriage that he instantly produces a dram of brandy which rallies the blood to the heart and generally prevails.

Dickens died five years later, on the anniversary of the Staplehurst crash.

TO THIS HOUR I HAVE SUDDEN RUSHES OF TERROR, EVEN WHEN RIDING IN A HANSOM CAB, WHICH ARE PERFECTLY UNREASONABLE BUT QUITE INSURMOUNTABLE

The Comfort of the Pullman Coach of a late-Victorian Passenger Train by Harry Green (b. 1920).

and might even suffer exposure. Travelling inside carriages modelled on stagecoaches was not without its disadvantages either.

If the atmosphere in the small railway compartments became fetid it was tempting to open the window. However, passengers risked being enveloped in smuts and steam from the engine rather than enjoying the bracing fresh air they'd sought. The question of whether or not to open windows on stuffy journeys was often the cause of fall-outs between passengers.

Borrowing an American idea, railway companies began to introduce more luxuriously appointed accommodation with restaurant facilities from the 1880s. Self-contained sleepers were on the rails from 1873. However, the first train with a corridor running its entire length wasn't introduced until 1892. Until then lavatories were simply not available to third-class travellers, and guards were sometimes compelled to clamber along the outside of moving carriages.

An electric signal for railways invented by a Mr King of Paxton, Derbyshire and exhibited at the Crystal Palace Electric Exhibition in 1882. The signal worked by passing electricity through a series of signal-posts placed along a railway line, which would activate a clock hand demonstrating the length of time elapsing between trains.

There's no doubt that railways were evolving throughout the Victorian era, although not as quickly as some had hoped. Bigger engines were soon hauling more carriages, increasing capacity and cutting costs. However, they weren't going much faster than early models. Until braking issues were resolved drivers were compelled to travel more slowly than they might have liked.

Steel rails were introduced at a junction outside Derby by the Midland Rail Company in 1857. They replaced wrought-iron rails that were so subject to wear and tear they were renewed every six months. Ten years later the steel rails, which carried about 500 trains daily, were still in use. Steel rails enhanced the comfort of passengers, too, making for a smoother ride, and soon steel was used in bridges and in locomotive manufacture after industrial advances made it cheap to produce.

If safety was a worry for passengers it was an even bigger issue among railway staff. In the 1870s an average of 750 railway employees were killed each year as managers turned a blind eye to even basic safety measures. However, advances in signalling did mean there were fewer collisions.

In early railway history, signalling was rudimentary and varied between companies. Alarmingly, two red discs on a hand-held wand indicated the line was 'all clear' on the Great Western while the same signal meant 'danger' on the London & Birmingham. Although flag-wavers were placed along busy lines, most drivers could only hope to stop in time if they saw danger ahead and could muster sufficient braking power.

A system of 'blocks' was then devised. Tracks were divided into a series of blocks and a train was only given permission to proceed if the block ahead was empty. At first signallers allowed 10 minutes between trains – although they still could not be sure the line ahead was empty. The number of trains permitted to run was duly curtailed. Soon railway managers pushed for the time gap to be reduced and, as a consequence, the number of accidents increased again.

The introduction of the telegraph brought about automatic signalling, in use from the 1840s. Soon the rail-side flag waver was replaced by a signalman who had his own dedicated space at the station. Typically, a weather-boarded box stood in a commanding position at one end of the platform after the 1860s, and by 1900 there were more than 13,000 in use all over Britain.

When he was president of the Board of Trade in the mid-1840s, William Gladstone set about involving government in the higgledy-piggledy railway administration. However, railway companies were staunchly opposed to sharing profits and power with government and used their political leverage to scotch the plans. Eventually, the Railway Clearing House (RCH) was instituted in 1842 to sort out who owed what to whom as a result of long-distance railway journeys. Railway companies were invited to join, although the RCH didn't achieve national coverage for more than 20 years.

Mostly the RCH accountants relied on honesty. However, an army of some 500 men were regularly dispatched to sit at key railway junctions noting every carriage and wagon that went by. For such perilous work they were relatively poorly paid, out in all weathers, and worked 13-hour shifts. High standards were generally expected of railway workers, as the 1856 rulebook issued by the Taff Vale Company reveals.

> *Every person is to come on duty daily, clean in his person and clothes, shaved, and his shoes blacked.*
>
> *It is urgently requested every person … on Sunday and Holy Days, when he is not required on duty that he will attend a place of worship; as it will be the means of promotion when vacancies occur.*

EVERY PERSON IS TO COME ON DUTY DAILY, CLEAN IN HIS PERSON AND CLOTHES, SHAVED, AND HIS SHOES BLACKED

William Ewart Gladstone (1809–1898).

(In addition, singing, whistling, laughing and political activity were banned. No one could become a stationmaster unless he was married, and all staff had to salute officers and directors.)

The difference in gauges across the networks was another issue that needed a national view. Confusion reigned at Gloucester station, where the broad-gauge Great Western Railway met the rest. Every passenger, animal and pound of freight had to make a change. Brunel had proved his broad gauge was superior when it came to speed, hauling an 80-ton load at almost 60 mph on one occasion. But the Royal Commission established to investigate railway gauges thought there was more to the argument than mere speed. It was cheaper to make a broad gauge narrow than vice versa, and there was already much more standard gauge on the ground than broad. Consequently, standard gauge was required by law in all railways built after 1846,

'Gauge War': an 1845 cartoon by Angus Reach on the conflict between the broad gauge of the Great Western Railway and the narrow gauge.

THE BROAD GAUGE AND THE NARROW GAUGE.

although the Great Western Railway didn't say farewell to its wide lines entirely until 1892.

The railways gave impetus to ship building, dock trade, harbours, hotels and the leisure industry. They modified everyone's lives and existing businesses beyond recognition.

Immense changes were made in the postal system, for example. Before the advent of the railways letters and parcels were transported with comparative speed by horse-drawn mail coaches that travelled across the nation. However, it was soon apparent that trains would do the same job more quickly and soon mail coaches became second-rate alternatives to rail.

Enterprising postal chiefs soon realised that trains could not only deliver the mail more rapidly but could be used as travelling sorting offices to further speed the progress of the post. As early as 1838 mail was sorted in a moving train, on this occasion a carriage made from a converted horse-box, on the Grand Junction Railway. The idea was seized upon by other services and incorporated into trains. These were first known as Railway Post Offices and later as Travelling Post Offices or TPOs.

From 1852 the mail bag pick-up apparatus became a common sight at the side of main lines. Aboard the train, postal workers secured letters inside leather pouches weighing anything between 20 and 60 lb (9–27 kg). These were attached to hinged metal arms that could swing out through open doors, dangling the pouch a few feet above the ground and several feet from the side of the carriage. A robust line-side net then scooped up the pouch and the metal arm retracted. To pass mail to a train, a postman positioned similar leather pouches in nets by the tracks so an arm from the train could 'catch' them and toss them through the open doors into the train.

Today it seems an old-fashioned and inefficient way to exchange mail, yet the system was used throughout Victorian times and well into the twentieth century, until 1971.

Throughout the era, the volume of post continued to increase thanks to a uniformity imposed by Sir Rowland Hill, who published his pamphlet *Post Office Reform* in the year Victoria came to the throne. Hill saw that, thanks to the railways, mail could be distributed

with increasing ease across the nation. If the cost was kept low more people than ever would use the service, he predicted. By 1840 the Penny Black stamp, bearing the profile of Queen Victoria and snipped from a sheet with scissors, came into use. Within a few years the first Christmas card was sent. By the mid-1850s pillar boxes were appearing on street corners to aid the collection of mail, and by Victoria's death there were some 32,500 pillar boxes in the UK. Without the age of the train, it is difficult to see how the postal service would have proliferated.

There were victims that fell to the age of rail, including stagecoaches, a sound road system built by Turnpike Trusts that eventually went out of the business, numerous ancient sites ploughed up for rail beds, and the rural economy. But the trains brought progress apace. At Victoria's death, *The Times* contemplated the achievements won in her lifetime and in its first breath addressed the expansion of railways.

> *Viewing that reign in its incidents, what a chronicle it offers of great national achievement, startling inventions and progress in every direction.*
>
> *The first railway was constructed before Victoria came to the throne but the universal development in the appliance of steam and electricity took place in her time and it profoundly altered the conditions of political and social life.*

More than just a method of getting from A to B, the railways imbued the country from top to bottom with a sense of 'can do' that had for the most part been sorely lacking. No aspect of life was left untouched by trains, and the interwoven railway lines that webbed the country were springboards for still more social and economic progress in the twentieth century.

There's an argument to say that every train trip of the nineteenth century fell into the bracket of 'great Victorian railway journeys', for there were sights, sounds, smells and tastes that are rarely re-created today. More than a series of sentimental journeys in pastoral England, the following five expeditions echo the eccentricities and evolution of Victorian railways, glimpsing an age of lines and locomotives the legacy of which is still evident today.

PREVIOUS PAGE: **A parcel-sorting centre on the Great Western Railway.**

RIGHT: **At Fenchurch Street Station.**

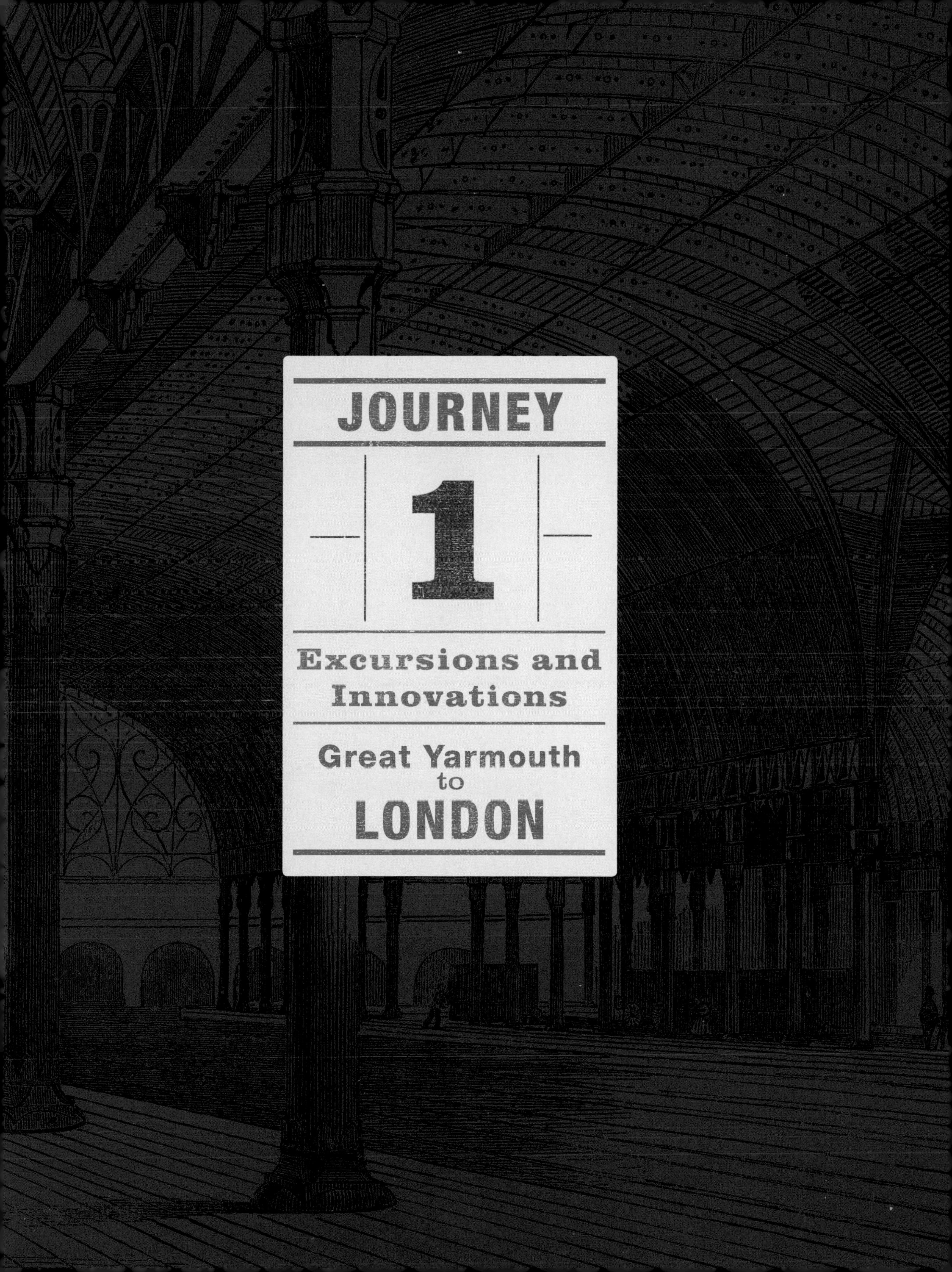
JOURNEY
1
Excursions and
Innovations
Great Yarmouth
to
LONDON

In Victorian times Great Yarmouth was fabled for two commodities: herring and holidays. The fishing industry was established long before the era began, peopled in part by Scottish fishermen who had sailed down with their families to live on the sandy promontory and exploit the shimmering shoals of the North Sea. Fish were then salted, barrelled and sent across the country.

Initially, the town was bounded by walls and fishermen lived cheek by jowl with one another, crammed into streets known as The Rows. For decades it was The Rows that gave Yarmouth its defining features, and they expanded to fill every available inch. Several wider roads ran roughly parallel with the waterfront. Narrower passages extended from those roads at right angles, creating a medieval grid that incorporated housing for rich and poor alike.

The Rows were so narrow that a law was passed to ensure doors opened inwards rather than outwards, to avoid injury to passers-by. Daylight and privacy were at a premium for the inhabitants. Drains that acted as open sewers ran down The Rows, with good community health dependent on prevailing winds and driving rain to drive the steady outpouring of sewage into the sea.

Author Charles Dickens was struck by the bunched-up quaintness:

> *A Row is a long, narrow lane or alley quite straight, or as nearly as maybe, with houses on each side, both of which you can sometimes touch at once with the fingertips of each hand, by stretching out your arms to their full extent.*
>
> *Now and then the houses overhang and even join above your head, converting the row so far into a sort of tunnel or tubular passage. Many picturesque old bits of domestic architecture are to be found among the rows. In some rows there is little more than a blank wall for the double boundary. In others the houses retreat into tiny square courts where washing and clear starching was done.*

Eventually Yarmouth's population outgrew the confines of the thirteenth-century town walls and, led by the example of wealthy merchants, spilled over on to nearby land formed when the seaways silted up.

Bradshaw's guidebook mentions to the town's fishing industry, also making reference to The Rows:

PREVIOUS PAGE: The Great Western Railway Terminus at Paddington Station.

RIGHT: Great Yarmouth, Row Number 60, 1908.

OVERLEAF: A steam train passes Weybourne windmill on the North Norfolk Railway connecting Sheringham with Holt.

61572

The Jetty, Great Yarmouth

The old town contains about 150 streets or passages, locally called rows, extending from east to west, in which many remains of antiquity may still be traced ... the inhabitants are chiefly engaged in the mackerel, herring or deep sea fisheries which are here prosecuted to a very great extent with much success.

Yet it makes little reference to the holiday trade which was by then beginning to boom. Great Yarmouth had long been a destination for a few well-heeled tourists who enjoyed the fresh air and the perceived benefits of sea water.

It was the arrival of trains that fired up the holiday trade, with trippers coming from London and other cities to sample the delights

IT WAS THE ARRIVAL OF TRAINS THAT FIRED UP THE HOLIDAY TRADE, WITH TRIPPERS COMING FROM LONDON AND OTHER CITIES TO SAMPLE THE DELIGHTS OF THE EAST COAST

of the east coast. Without the onset of train travel, it's doubtful that the national passion for a trip to the seaside would ever have taken root, for travel by coach was slow and expensive by comparison. The town's first station, known as Yarmouth Vauxhall, opened in 1844, and so popular was Great Yarmouth as a destination that one estimate insists more than 80,000 people visited the resort just two years after that station opened.

Great Yarmouth was once served by four separate train lines, and a clutch of town centre stations and no fewer than 17 other stations were spread around the borough. It was such a popular destination that the Great Eastern Railway produced postcards featuring views of Great Yarmouth to sell to its passengers.

When a suspension bridge collapsed on 2 May 1845, killing 79, the dead surely included some of the new influx of tourists. People had gathered on the bridge to watch a clown in a barrel being towed down the River Bure by a team of geese. As the barrel passed under the bridge they rushed for the other side to catch more of the spectacle, causing supporting chains to snap. Scores of people, mostly women and children, were hurled into the river and local men took to their boats to save them.

According to an account in the *Norwich Gazette*, tragedy on a far greater scale was averted:

> *It can be easily imagined that a mass of people thus precipitated into water, five feet deep, would have but a small chance of saving themselves; and but for the prompt assistance which was afforded, few, very few, would have escaped. Boats and wherries were immediately in motion and from 20 to 30 with gallant crews, were soon among the drowning people, picking them up with wonderful rapidity. Many were put on the shore in their wet clothes who went directly home, and no account was taken of the number thus saved.*

The tombstone of nine-year-old bridge disaster victim Thomas Beloe, in nearby St Nicholas' Churchyard, depicts the tragedy. In fact saving lives became something of a theme for Great Yarmouth, with local boat-builder James Beeching winning the 100 guinea first prize in an 1851 competition to find the best self-righting lifeboat.

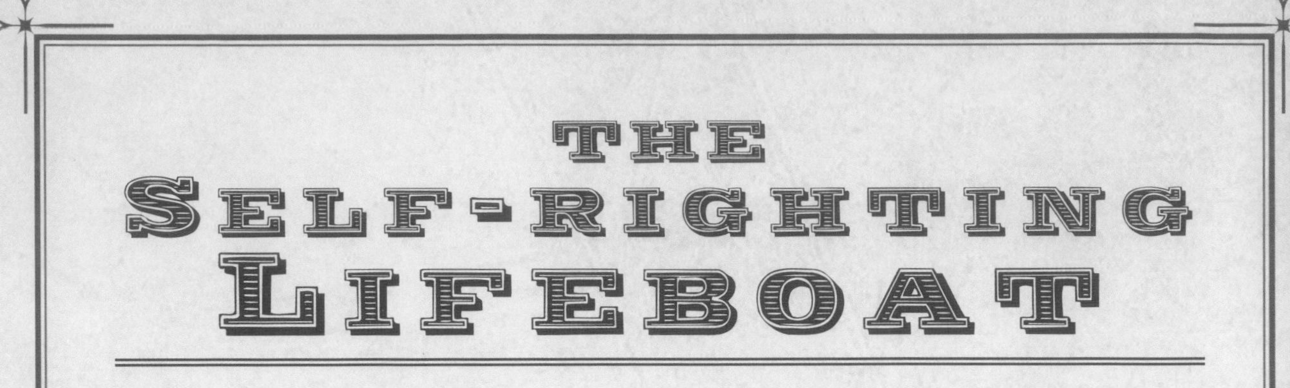

The Self-Righting Lifeboat

Lifeboat design was still in its infancy in the Victorian era and the 1851 competition was launched to design a new and better boat. It had several stated aims. Lifeboats of the future needed to be lighter in construction than previous models so that they could more easily be launched from the beach. They also needed to be cheaper to make so that more could be produced. With such generous prize money on offer the competition attracted 280 entries from across Britain, Europe and even the USA.

Following adjustments, and with inspiration taken from other designs submitted for judging, the Beeching lifeboat became the basis of the longstanding Norfolk and Suffolk class of boats. Throughout the second half of the nineteenth century the design was improved, but the Beeching boat's enduring feature was its buoyancy, with air-filled cases at the bow and stern and cork cladding. It effectively discharged the seawater which frequently swamped small, open boats through valved tubes, and an iron keel acted as ballast. It was stable, self-righting, fast, robust and comparatively roomy. Boats like this saved countless hundreds of lives during the remainder of the century.

A self-righting boat like Beeching's was popular with lifeboat men. Analysis of the number of capsizes between 1852 and 1874 showed their instincts were probably right. In that time, 35 self-righters rolled with the loss of 25 men out of a total of 401. At the same time 8 non-self-righters capsized, killing 87 men out of 140.

A lifeboat rests on its carriage, c. 1880.

However, a train from Great Yarmouth heading for Norwich was involved in a night-time collision on 10 September 1874 in which 25 people died and 50 were injured. It occurred after a signalling error had allowed a 14-coach mail train to rush headlong on a single track into the 13-coach passenger train from Yarmouth. Although the fronts of both trains were smashed to smithereens, the rear coaches were left relatively unscathed. One account of the accident ends with an odd incident that perplexed everyone who witnessed the wreckage:

George Bidder (1806–1868).

It would be difficult to conceive of a more violent collision … yet it is said that two gentlemen in the last carriage of one of the trains, finding it at a sudden standstill close to the place to which they were going, supposed it had stopped for some unimportant cause and concluded to take advantage of a happy chance which left them almost at the doors of their homes. They accordingly got out and hurried away in the rain, learning only the next morning of the catastrophe in which they had been unconscious participants.

The Great Eastern main line from Yarmouth heads to Reedham, distinguished by one of four swing bridges in the area. This bridge across the River Yare, and the one at Somerleyton – on the branch line that connects Reedham with Lowestoft – spanning the River Waveney were financed by Sir Samuel Morton Peto, entrepreneur and engineering enthusiast.

Both bridges are made from a stout collection of wrought iron, brick, cast steel and timber. When it is in place for trains, the bridge ends rest on piers by the river banks. If it is open for river traffic then the bridge pivots on a central pier using cast steel wheels with a diameter of 16 inches. The load of the open bridge is shouldered by two truss girders.

Even today the bridges are an object of wonder. The man who built the bridges, George Bidder, was equally remarkable. The son of a Devon stonemason, his natural ability with maths manifested itself before he could read or write and his father had him perform in shows around the country for money, under the title of 'a calculating boy'. Fortunately, his potential was spotted by two benefactors, who ultimately paid for his education. In adulthood he teamed up with the

The Reedham railway swing bridge crossing the River Yare.

great Robert Stephenson to work on major railway projects at home and abroad. Perhaps his proudest achievement was to build London's Victoria docks.

The man who financed the swing bridges has a story that perhaps even exceeds that of Bidder. Sir Samuel Morton Peto was born in Woking, Surrey, to a tenant farmer. After two years at boarding school he was made an apprentice to his builder uncle, Henry Peto. In 1830 he took over the business with his cousin Thomas Grissell, and together they changed the landscape of London by building the Houses of Parliament, Trafalgar Square and Nelson's Column, among other landmarks. The business then became involved in building railways.

After he bought Somerleyton Hall in 1844, Peto invested heavily in the area, fashioning Lowestoft into a thriving port and town. He built the railway line to it from Reedham, which opened in 1847 after some two years in construction.

ABOVE: **Somerleyton Hall, the home of Samuel Peto.**

OVERLEAF: **The inauguration of the Great Industrial Exhibition of 1851.**

However, his partner Grissell was becoming nervous about what he perceived as reckless risks taken by Peto in pursuit of railway contracts. The partnership was dissolved and Peto began business anew with his brother-in-law Edward Betts in 1846. They also worked with engineer Thomas Brassey, a millionaire railway builder and civil engineer credited with an enormous number of projects. Previously Brassey had worked with George Stephenson and his acolyte Joseph Locke, and by the time he died Brassey had built a sixth of the railways in Britain and half of those in France.

The trio of Peto, Betts and Brassey built numerous railways at home and abroad. Peto earned the gratitude of Prince Albert by ensuring there were suitable rail links to the Great Exhibition at Crystal Palace in 1851. However, one of the most significant contributions Peto – with Betts and Brassey – made to history was to build a rail link in the Crimea, where Britain was at war with Russia.

INDIA

In a conflict ignited by Russian occupation of Turkish territories, British hopes of a swift victory were confounded by climate and disease. However, Britain and her allies got back on the front foot with the first strategic use of railways, built and paid for by Peto. A railway line to ferry men and supplies to the front line was in operation by 1855. Five months later the British target, Sebastopol, had fallen.

That same year Peto was given a baronetcy, and for 20 years he was an MP. But in 1866 his riskier ventures caught up with him as the frenzied speculation in railway building known as 'railway mania' reached its third crescendo and brought down a bank, Overend, Gurney & Company, to which Peto was deeply committed. With the bank entering liquidation in 1866 owing about £11 million, he was declared bankrupt. Peto moved to Budapest, hoping to spark railway building there, but he met with no success. He moved back to Britain but died in obscurity in 1889.

Peto had shouldered a lot of the small East Anglian lines into existence. In 1862 many of the small, east coast companies, including the Norfolk Railway, Eastern Union Railway, East Norfolk Railway, Newmarket & Chesterford Railway, Harwich Railway and the East Suffolk Railway, were mopped up by the Great Eastern Railway, along with the more major Eastern Counties Railway. Although they were now officially all one company, it took years for the competitive habit between lines to fall by the wayside.

The branch line to Lowestoft was not the only one to extend from the railway that linked Great Yarmouth to Ipswich. As Lowestoft's fortunes increased, so Southwold further down the coast became the poorer. A lack of railway line was clearly a factor in any future prosperity the town might enjoy, so local people clubbed together to buy shares in the Southwold Railway Company that would join the main line at Halesworth.

After protracted negotiations, a 3-foot gauge was chosen for the route, which had a single track that ran for nearly nine miles. It opened on 24 September 1879. The locomotives that used the line were limited to a speed of 16 mph and before long it was quicker to cycle there than take the train. Coupled with a reputation for unreliability and a laissez-faire attitude among station staff, the line became something of a laughing stock and the subject of jokey postcards.

But there was plenty to recommend Southwold, including an attractive North Sea coastline and its proximity to Dunwich, the medieval city that was reclaimed by the sea after a series of violent storms. Victorian curiosity was piqued about the city, which by all accounts at one time had 52 churches, city walls, a Royal palace and a mint. In 1900 the railway carried 10,000 passengers, 90,000 tons of minerals and 600 tons of merchandise. It finally closed in 1929 in the face of fierce competition from buses.

Another branch line further south at Ipswich led to Felixstowe, today the nation's biggest commercial port, which had its fate and fortunes defined by one man.

George Tomline was an enormously wealthy MP who made his home at nearby Orwell Park, named for the River Orwell, in Suffolk. After rebuilding the house he furnished it with fine art, an extensive library and even an observatory. He was known as 'Colonel Tomline' but the title was the result of a loose association with a regiment rather than distinguished military service.

Tomline conceived the plan for a railway line that would leave the main Great Eastern Railway at Westerfield and would head for Felixstowe via Orwell Park. On the face of it there was naked self-interest in having his own railway station. However, Tomline maintained his motive was to provide work for hard-pressed local people.

The odds were stacked in his favour from the outset. Following a concerted campaign of purchasing he already owned most of the land needed for the route. When it applied for parliamentary approval in 1875, his company was called the Felixstowe Railway & Pier Company. Within two years (and at a cost of £14,000) the line had opened with three locomotives, 19 passenger carriages and 15 goods wagons on the line. At Felixstowe he built a beach-side station, which was not only on land he owned but was as far away as possible from the Ordnance Hotel, owned by Ipswich brewery magnate John Chevallier Cobbold – a man Tomline apparently detested.

Within two years of its opening, the running of the line was given over to the Great Eastern Railway – but it wasn't the end of the story

PREVIOUS PAGE: British navvies commence work on the construction of a railway line between Balaclava and Sebastopol in the Crimea in 1855.

RIGHT: A steam train approaching Weybourne Station on the North Norfolk Railway.

65462

as far as Tomline and Felixstowe were concerned. In 1884 his company was renamed the Felixstowe Dock & Railway Company, having secured the necessary permissions for construction work that would provide moorings, warehousing and railway sidings. Although Tomline gave up his interest in the railway three years later he maintained a link with the dock, which finally opened in 1886, three years before his death. Since then it has grown beyond all expectation.

Felixstowe finally got a town-centre railway station in 1898, courtesy of the Great Eastern Railway.

Midway between Ipswich and Colchester, Suffolk gives way to Essex, although the slow pace of rural life remained the same. When the Great Eastern main line crossed the River Stour on the Essex and Suffolk border it bisected an area known today as Constable Country. It contains, of course, the vistas that inspired artist John Constable. Some of his most famous works, including *The Haywain* and *Flatford Mill*, were painted here near his boyhood home of East Bergholt in Suffolk.

Constable died in 1837, the year Victoria came to the throne. During his lifetime his paintings were more popular in France than ever they were in England. Both he and fellow artist J. M. W. Turner were lambasted by critics of the day for being safe and unadventurous in their work. But Constable insisted he would rather be a poor man in England than a rich one overseas and stayed to forge a living in the only way he knew how.

His inspiration was nature, and his pictures often betrayed the first intrusions of the Industrial Age into rural life. Although he didn't always live there, it was Suffolk scenes he was perpetually drawn to paint. 'I should paint my own places best,' he wrote. 'Painting is but another word for feeling.' An indisputably Romantic painter, his rich use of colour arguably laid the foundations for future trends in art.

The tallest Tudor gatehouse ever built lies further down the line, marking the half-way point between Colchester and Chelmsford. Aside from its architectural glory, Layer Marney Tower has two striking claims to fame. The first is that it was owned from 1835 by Quintin Dick, an MP made notorious by his practice of buying votes. Indeed,

LAYER MARNEY TOWER HAS TWO STRIKING CLAIMS TO FAME

RIGHT: **Layer Marney Tower, a Tudor palace damaged by the Great English Earthquake of 1884.**

there's some speculation that he spent more money bribing his constituents than any other MP of the era. The son of an Irish linen merchant, Dick spent a total of 43 years as an MP, representing five different constituencies.

The tower's second claim to fame is that it and surrounding buildings built during the reign of King Henry VIII were badly damaged in the Great English Earthquake of 1884, which had its epicentre near Colchester. Afterwards a report in *The Builder* magazine stated the author's belief that the attractive monument was beyond repair: 'The outlay needed to restore the tower to anything like a sound and habitable condition would be so large that the chance of the work ever being done appears remote indeed.'

However, the tower was repaired, thanks to the efforts of the then owners, brother and sister Alfred and Kezia Peache, who re-floored and re-roofed the gatehouse, and created the garden to the south of the tower. Layer Marney Tower was one of an estimated 1,200 buildings damaged by the earthquake, which struck on 22 April and measured 4.6 on the Richter scale. There were conflicting reports about a possible death toll, ranging from none to five. The earthquake sent waves crashing on to the coastline where numerous small boats were destroyed.

From the main east coast line it eventually became possible to forge across country by branch line to Southend. It wasn't the earliest line built to the resort, however, nor would it be the busiest. Contractors Brassey, Betts and Peto built the first railway into Southend from London, although plans to site the station at the town's pier head were vetoed on grounds of nuisance. It was the last stop on a line that went via Tilbury and Forest Gate to either Bishopsgate or Fenchurch Street. Primarily managed by the London, Tilbury & Southend Railway Company, the line was known locally as the LTS.

After the railway was opened there was extensive development in the town, providing houses large and small at Clifftown. Samuel Morton Peto was once again a moving force in the plans. The homes were completed in 1870 and, a decade later, a newly designed tank engine went into operation on the LTS which could haul more people at faster speeds than ever before. For the first time people could live in Southend while working in London with ease, thanks to the train. Thus Southend became an early commuter town, as well as being the closest resort to London.

In 1862 at Crewe, John Ramsbottom, chief engineer of the London & North Western Railway, proved the versatility of an 18-inch gauge for industrial trains, which could run not only up to but into warehouses. Eventually the gauge at Waltham Abbey was changed, so when production went into overdrive during the First World War the factory was at its most efficient.

Freight across the Great Eastern Railway was for years dominated by food. In addition to fish from the east coast there were vegetables – linking the fortunes of the railway company inextricably to the wealth of the harvest. There was also milk, which first travelled in churns hoisted into ventilated vans to keep it as fresh as it could be for thirsty city folk. This way the train service made a significant contribution to the health of the nation, supplying fresh food to cities at comparatively low costs.

In the same way (but in the opposite direction), railways carried newspapers fast and efficiently into rural areas, improving education and awareness everywhere in a way that was once confined to cities.

In 1847 the Eastern Counties Railway began to build a depot at Stratford where its locomotives were made. It was extended time and again throughout its history until it became a maze of track and workshops. In 1891, when it was under the aegis of the Great Eastern Railway, a new record was set there for building a locomotive. It took just nine hours and 47 minutes to produce a tender engine from scratch, complete with coat of grey primer. As a sign of the frantic railway times, the locomotive was dispatched immediately on coal runs, and covered 36,000 miles before returning to Stratford for its final coat of paint. Its working life lasted for 40 years and it ran through 1,127,000 miles before being scrapped.

When *Bradshaw's* was written in 1866, the terminus of the Great Eastern line was Bishopsgate in Shoreditch. The guidebook calls it 'one of the handsomest (externally) in London'. It was opened in 1840 by the Eastern Counties Railway and its name was changed from Shoreditch to Bishopsgate in 1847.

When Eastern Counties Railways amalgamated with other lines to form Great Eastern Railways, the new company found its two options for terminals – Bishopsgate and Fenchurch Street Stations – were not sufficiently large and set about building Liverpool Street Station and its approach tunnel, which opened in 1874.

An engraving of Bishopsgate Street by Gustav Doré, 1872.

IT TOOK JUST NINE HOURS AND 47 MINUTES TO PRODUCE A TENDER ENGINE FROM SCRATCH, COMPLETE WITH COAT OF GREY PRIMER

DOUBLE
STOUT
PERKINS
MANUFACTORY
SIR PAUL PINDAR
STOUT HOUSE

Railway carriages at Weybourne Station on the North Norfolk Railway.

Nowhere in Britain has the railway map changed more than in London, not least due to the Blitz in the Second World War. In 1866 it was possible to jump on a North London line train at Fenchurch Street or Bow, within moments of getting off a Great Eastern line train.

This is a route that became infamous in 1864 for being the scene of Britain's first train murder. The victim was 69-year-old Thomas Briggs, a senior clerk at the City bank Messrs Robarts, Curtis & Co. On Saturday 9 July he had worked as he always did until 3 p.m. and then visited a niece in Peckham before making his way home by train.

No one knows just what happened in the first-class carriage of the 9.50 p.m. Fenchurch Street service. It was, in common with many other carriages, sealed off from other travellers. There were six seats, three on each side, and two doors in a design reminiscent of stagecoaches. Subsequent passengers found the empty seats covered in blood and an abandoned bag, stick and hat. Almost simultaneously, a train driver travelling in the other direction saw a body lying between the tracks. After he raised the alarm the badly injured Mr Briggs was carried to a nearby tavern but he died later from severe head injuries.

There was a public outcry at the killing, although crimes like theft and even assault had been carried out on trains almost since their inception. Now, however, a sense of peril accompanied train travel as never before.

At first there seemed little for detectives to go on. Mr Briggs's family identified the stick and bag as his but the hat was not, and his own hat was missing. Cash was left in his pocket but his gold watch and chain were gone.

A wave of scandalised press coverage yielded the first clue. It alerted a London silversmith, appropriately called John Death, who told police he had been asked to swap Mr Briggs's watch chain for another, and described the customer making the request. Later, a Hansom driver confirmed that a box with the name Death written on it was at his house, brought there by a German tailor, Franz Muller, who had been engaged to his daughter. The Hansom driver obligingly produced a photo of Muller and the silversmith confirmed him to be the watch-chain man.

Before a warrant could be issued for his arrest, Muller had boarded the sailing ship *Victoria* bound for New York in anticipation of a new life in America. Detective Inspector Richard Tanner, along with his material witnesses, soon booked tickets aboard the steamship *City of*

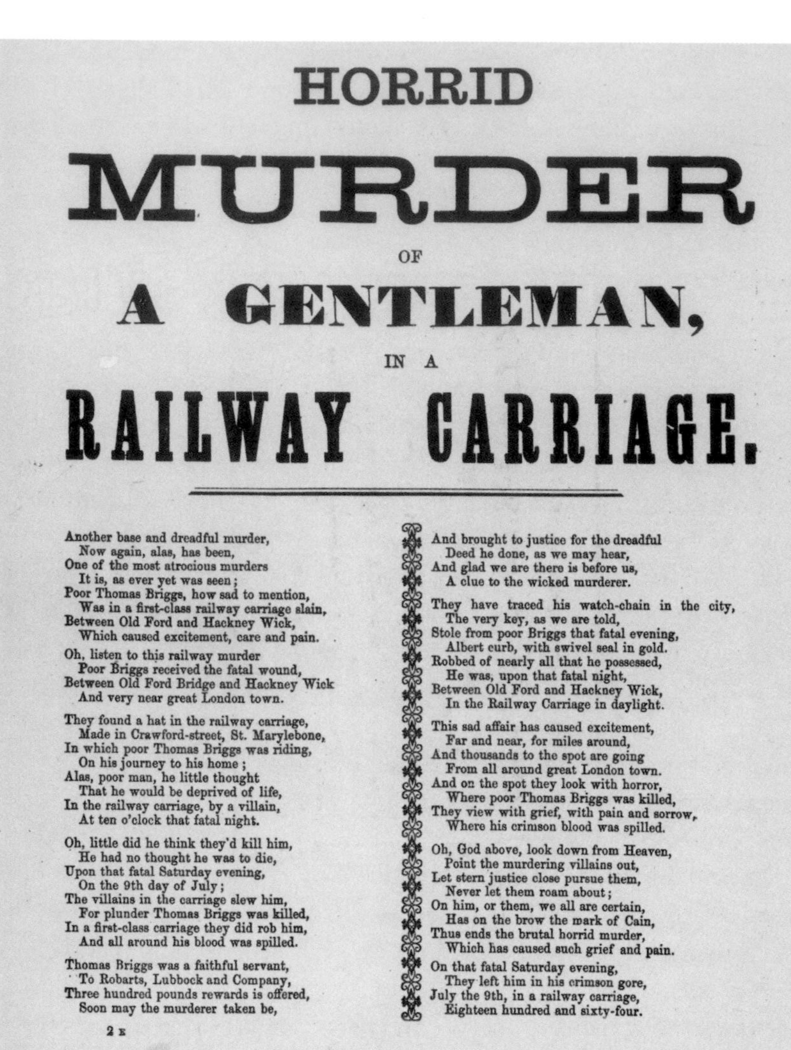
HORRID
MURDER
OF
A GENTLEMAN,
IN A
RAILWAY CARRIAGE.

Another base and dreadful murder,
Now again, alas, has been,
One of the most atrocious murders
It is, as ever yet was seen;
Poor Thomas Briggs, how sad to mention,
Was in a first-class railway carriage slain,
Between Old Ford and Hackney Wick,
Which caused excitement, care and pain.

Oh, listen to this railway murder
Poor Briggs received the fatal wound,
Between Old Ford Bridge and Hackney Wick
And very near great London town.

They found a hat in the railway carriage,
Made in Crawford-street, St. Marylebone,
In which poor Thomas Briggs was riding,
On his journey to his home;
Alas, poor man, he little thought
That he would be deprived of life,
In the railway carriage, by a villain,
At ten o'clock that fatal night.

Oh, little did he think they'd kill him,
He had no thought he was to die,
Upon that fatal Saturday evening,
On the 9th day of July;
The villains in the carriage slew him,
For plunder Thomas Briggs was killed,
In a first-class carriage they did rob him,
And all around his blood was spilled.

Thomas Briggs was a faithful servant,
To Robarts, Lubbock and Company,
Three hundred pounds rewards is offered,
Soon may the murderer taken be,

2 E

And brought to justice for the dreadful
Deed he done, as we may hear,
And glad we are there is before us,
A clue to the wicked murderer.

They have traced his watch-chain in the city,
The very key, as we are told,
Stole from poor Briggs that fatal evening,
Albert curb, with swivel seal in gold.
Robbed of nearly all that he possessed,
He was, upon that fatal night,
Between Old Ford and Hackney Wick,
In the Railway Carriage in daylight.

This sad affair has caused excitement,
Far and near, for miles around,
And thousands to the spot are going
From all around great London town.
And on the spot they look with horror,
Where poor Thomas Briggs was killed,
They view with grief, with pain and sorrow,
Where his crimson blood was spilled.

Oh, God above, look down from Heaven,
Point the murdering villains out,
Let stern justice close pursue them,
Never let them roam about;
On him, or them, we all are certain,
Has on the brow the mark of Cain,
Thus ends the brutal horrid murder,
Which has caused such grief and pain.

On that fatal Saturday evening,
They left him in his crimson gore,
July the 9th, in a railway carriage,
Eighteen hundred and sixty-four.

A report, taking the form of verse, on the murder of Thomas Briggs in a railway carriage on 9 July 1864.

Manchester, easily beating the *Victoria* to its destination. In fact, the Metropolitan Police party had to wait four weeks for it to catch up. When the police finally approached Muller on the dockside he asked, 'What's the matter?'

A swift search established he was in possession of Mr Briggs's watch and remodelled hat. At the time, relations between Britain and America – torn by civil war – were strained. Nonetheless, a judge agreed to extradite Muller and he was soon brought back to England.

Muller maintained his innocence throughout his Old Bailey trial and claimed he bought the watch and hat on the London dockside. He was small, mild-mannered and apparently lacked a motive. There were also witnesses to say Mr Briggs was seated with not one but two men on the night he was killed. But the jury took just 15 minutes to find Muller guilty.

Despite pleas for clemency from the Prussian King Wilhelm I, Muller was publicly hanged at Newgate Prison just four months after the crime. Later the prison chaplain claimed his final words were 'I did it'. Still, his death nearly resulted in a riot, with many Londoners filled with doubt about the verdict.

Franz Muller, a German tailor found guilty of the murder of Thomas Briggs and hanged ouside Newgate prison in 1864.

The savage killing of Thomas Briggs resulted in new legislation, introduced in 1868, which made communication cords compulsory on trains. Although open carriages were still viewed unfavourably it was felt Mr Briggs's life could have been saved if the train driver only knew he had been in difficulties.

In 1897 an American journalist, Stephen Crane, travelled on the Scotch Express between London and Glasgow, and revealed that, some 30 years after the death of Mr Briggs, communication cords were causing unforeseen difficulties. The problem arose when dining cars came into use and shared the same alarm system, causing confusion. He wrote:

> *...if one rings for tea, the guard comes to interrupt the murder and that if one is being murdered, the attendant appears with tea. At any rate, the guard was forever being called from his reports and his comfortable seat in the forward end of the luggage van by thrilling alarms. He often prowled the length of the train with hardihood and determination, merely to meet a request for a sandwich.*

Moved by Mrs Briggs's plight, spy holes were drilled in carriage partitions by some train companies, and became known as 'Muller lights'. Bizarrely, Mr Briggs's reshaped hat became something of a fashion item.

The North London Railway established a depot on a 10-acre site at Bow in 1853 where it built and repaired its own locomotives for the remainder of the Victorian era. At the time East London was assuming a reputation for poverty and moral decline. Most families

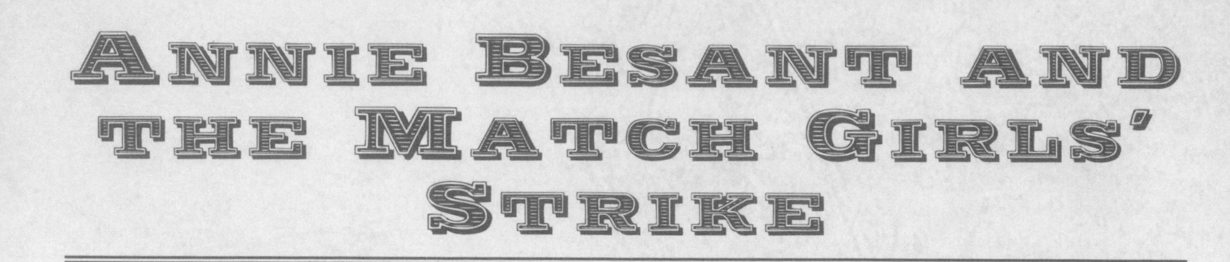

Annie Besant and the Match Girls' Strike

Annie Besant, a campaigning social reformer, decided to investigate claims about the ill treatment of match workers at Bryant & May's factory, dangerous conditions and the company's system of fines for petty misdemeanours. She reported what she found after interviewing some of the 'match girls' in her journal, *The Link*, in June 1888 under the headline 'WHITE SLAVERY IN LONDON'.

Annie Besant (1847–1933).

The splendid salary of 4s. is subject to deductions in the shape of fines; if the feet are dirty, or the ground under the bench is left untidy, a fine of 3d. is inflicted; for putting 'burnts' — matches that have caught fire during the work — on the bench 1s. has been forfeited, and one unhappy girl was once fined 2s. 6d for some unknown crime. If a girl leaves four or five matches on her bench when she goes for a fresh 'frame' she is fined 3d., and in some departments a fine of 3d. is inflicted for talking. If a girl is late she is shut out for 'half the day', that is for the morning six hours, and 5d. is deducted out of her day's 8d. One girl was fined 1s. for letting the web twist round a machine in the endeavour to save her fingers from being cut, and was sharply told to take care of the machine, 'never mind your fingers'. Another, who carried out the instructions and lost a finger thereby, was left unsupported while she was helpless. The wage covers the duty of submitting to an occasional blow from a foreman; one, who appears to be a gentleman of variable temper, 'clouts' them 'when he is mad'.

Besant was gravely concerned that working with phosphorus used at the factory – already banned in Sweden and the USA – was causing cancer. (The British government refused a ban on the grounds it would restrain free trade.)

The company's owners, Quakers Francis May and William Bryant, were furious, branding Besant's newspaper claims as lies and hounding those they believed were responsible for talking to her.

When the factory owners forced their employees to sign a statement saying they were happy with working conditions, 1,400 women went on strike with Besant at their head. Their campaign attracted some high-level support, including from the *Pall Mall Gazette*, Catherine

The Bryant & May match factory.

Booth of the Salvation Army and the writer George Bernard Shaw. However, they were also lambasted by others, including *The Times*.

Determined to beat the bosses, the strikers organised themselves as never before. There were marches in both the east and west end of London. There was a strike fund, with each contribution listed in an accounts book. For the first time the London Trades Council – formed in 1860 to represent skilled workers – lent its support, donating £20 to the strike fund and offered to mediate in talks.

A strike headquarters set up in Bow Road to coordinate action and maintain a register of everyone involved. The Strike Register reveals many of the women and girls were of Irish extraction and lived close to one another in nearby slums. Typically, the Irish already felt under attack by the British and British attitudes, and were more inclined to confront the Establishment than many English workers at the time.

After three weeks the company agreed to end the hated fines' system. The strikers were triumphant and infant union movements nationwide were given a boost.

On 27 July 1888 the inaugural meeting of the Union of Women Match Makers was held, with Besant elected as the first secretary. With money left over from the strike fund – as well as the profits from a benefit show held at a London theatre – the union found itself premises and enrolled 666 women. Before the year was out it became known as the Matchmakers' Union. Its story was short-lived as it folded in 1903, but its galvanising effect on the union movement continued for years afterwards.

Moreover, the Salvation Army went on to open its own match factory in East London, using a less harmful phosphorus and paying twice as much as Bryant & May. Bad publicity continued for the company, until in 1901 it announced an end to the use of harmful yellow phosphorus in its production process.

L.99

lived in single rooms in largely insanitary conditions. Once, the area was the domain of weavers and their families. Now their cloth-making skills were largely obsolete, although a couple of silk factories remained.

The strike by the match girls was not the only East End story to hit the headlines at the time. Between April 1888 and February 1891 11 women were murdered and mutilated by a man who became known as 'Jack the Ripper'. Despite a massive operation the police failed even to arrest, let alone convict, anyone for the crime.

Staff at the Bow Infirmary Asylum, which stood opposite the Bryant & May factory, felt sure one of their patients, an East European immigrant butcher called Jacob Isenschmid, was the culprit. He had been released from the asylum in 1887, apparently cured. After the fourth murder he was seen with blood on his clothes

OPPOSITE: **A steam train on the North Norfolk Railway.**

BELOW: **Horse-drawn trams and wagons outside the City of London Infirmary.**

in a pub close to the scene of the murder. Asylum staff contacted the police but, despite an interview, there was no evidence against him and he remained at large, although he does not appear on present-day lists of suspects.

London's transport systems were changing dramatically. There were new terminals built, usually in grand fashion, on the outskirts of the city centre to receive trains from all corners of the country. But for a while these stood in awkward isolation, although an ever-increasing number of lines were bolted on to the company or network they served, at the expense of London housing. Congestion on London's roads as travellers went between one railway line and the next intensified.

In writing *Dombey and Son*, which appeared in instalments between 1846 and 1848, Dickens described railway building in London.

> *Houses were knocked down; streets broken through and stopped; deep pits and trenches dug in the ground; enormous heaps of earth and clay thrown up; buildings that were undermined and shaking, propped up with great beams of wood ... Hot springs and fiery eruptions, the usual attendance upon earthquakes, lent their contribution of confusion to the scene...*
>
> *In short, the yet unfinished and unopened railroad was in progress; and, from the very core of all this dire disorder, trailed smoothly away upon its mighty course of civilization and improvement.*

In 1863 the notion of one mighty central terminus for all the capital's railways was once again rejected by a House of Lords Select Committee, anxious that no more housing in an already overcrowded city should be sacrificed for the sake of the railways. The following excerpt from *Hansard* reveals the extent of railway schemes laid before the Select Committee the following year.

> *We found that those schemes were of vast magnitude for so limited an area as the metropolitan district. The new railways proposed to be constructed within that area extended over a length of 174 miles in the aggregate, and involved the raising of capital to the amount of about £44,000,000. It was, of course, impossible all that mileage could be*

constructed, or all that capital expended for metropolitan railways, because many of those schemes were necessarily competing schemes. At the same time, my Lords, it must be confessed that there was sufficient cause for considerable alarm among the holders of property in the metropolis, and much reason to apprehend that, if any large number of these lines were sanctioned, the traffic of many important public thoroughfares would be seriously interfered with during the construction of those works. Those schemes, as they came before us, included the construction of no less than four new railway bridges across the Thames, two of them – and these of a very large size – being intended to cross the river below London Bridge.

PREVIOUS PAGE:
Construction of the Arches of St Pancras Station's Cellars, London, **by J.B. Pyne, c. 1867.**

There was by now an undercurrent of public feeling against the railways as they deposited viaducts, tracks and tunnels at will, altering the complexion of the capital forever. London, perhaps more than any other city, was almost entirely remodelled by the converging transport companies.

As early as 1864 the satirical magazine *Punch* asked plaintively: 'Are there no means of averting the imminent destruction of the little beauty that our capital possesses?' The article went on to say that, given the railway frenzy existing at the time, St Paul's Cathedral might just as well become a railway station.

The graveyard at St Pancras was removed for the sake of the railway. A coaching house that escaped the Great Fire of London in 1666 – in which about 13,500 homes and 87 parish churches were razed to the ground – lying in the shadow of St Paul's was destroyed in 1875 to make way for lines and stations. Hundreds more buildings were flattened to make way for tracks, including Sir Paul Pindar's house in Bishopsgate. Pindar, an ambassador to the Ottoman court for James I, owned a fine house with one of the most distinctive frontages in Victorian London that likewise escaped the Great Fire. In 1890 the house's distinctive Jacobean façade was dismantled in favour of an extension to Liverpool Street Station. Fortunately, the wooden structure found a new home in the Victoria and Albert Museum.

THE GRAVEYARD AT ST PANCRAS WAS REMOVED FOR THE SAKE OF THE RAILWAY

London's population was being squeezed into its outer reaches. Those houses that remained were smeared with smoke as steam trains brought dense and eerie pollution into the city. Only the very rich could resist the onward march of the railways.

CHRONOLOGY OF THE LONDON TERMINALS

1836 London Bridge Station was built in primitive form for the London & Greenwich Railway and was soon subject to a rebuild.

1837 Euston, operated by the London & North Western Railway.

1838 Paddington, still bearing the hallmarks of its designer Brunel, built to receive Great Western Railway services.

1841 Fenchurch Street, the smallest of the railway terminals in London, originally constructed for the London & Blackwall Railway, and rebuilt 13 years later in time to accommodate the London, Tilbury & Southend Railway. It was the site of the first station bookstall.

1848 Waterloo Bridge Station, as it was called, opened after being linked to the busy outer-city satellite at Nine Elms for the London & South Western Railway.

1852 King's Cross opened for the Great Northern Railway on the site of smallpox and fever hospitals. It was designed by Lewis Cubitt along remarkably simple lines save for an Italianate clock turret. A hotel was built to accompany the station and opened two years later.

1858 Victoria, named for the nearby street, was the home of London, Brighton & South Coast Railway trains, although it was soon popular with other companies.

1864 Charing Cross, arguably the only London station to breach the West End, opened with six wooden platforms for what was initially a limited service to Greenwich and mid-Kent.

1866 Moorgate came into being in an extension to the Metropolitan Line and only became a main-line terminus in 1900.

1868 St Pancras was built by the Midland Railway after it found King's Cross too expensive. It became remarkable for the railway hotel's vast Gothic frontage.

1874 Liverpool Street was built to replace Bishopsgate Station, being closer to the city centre and more user friendly.

1899 Marylebone was home to the final main line to enter London, the Great Central, but plans by chairman Sir Edward Watkin to continue expansion with a channel tunnel were never realized.

Railway company managers were powerful people but some left a more distinguished legacy than others.

Sir James Allport spent a career in railways, ending up as the boss of Midland Railways for 27 years, excepting a short spell spent at a shipyard in Jarrow. He was also instrumental in forming the Railway Clearing House, which managed payments between different companies to cover journeys spanning several networks. After his retirement as manager in 1880 he became a director of the company.

Under his leadership, Midland Railway services expanded and the grand station at St Pancras was opened. But he is best remembered for transforming the journeys of third-class passengers. He was the first to realise that, rather than being a hindrance to the railway company, third-class passengers were in fact a valuable asset.

Accordingly, he made third-class carriages much more comfortable and, from 1872, included third-class carriages on every train, charging passengers a penny per mile for a journey. When some angry passengers

PREVIOUS PAGE: *Charing Cross Station*, London, c. 1864, a coloured chromolithograph by the Kell brothers. The station was designed by John Hawkshaw and was the London terminus of the South Eastern Railway.

BELOW: *Seats for Five Persons* by Abraham Solomon (1824–1862).

RATHER THAN BEING A HINDRANCE TO THE RAILWAY COMPANY, THIRD-CLASS PASSENGERS WERE IN FACT A VALUABLE ASSET

boycotted Midland Services he scrapped second class, at the same time lowering first-class fares. The result was better revenues for the railway company and a more equitable system of travelling.

For his services to cheaper travel Allport was knighted in 1884. But in his later life it wasn't the gong at the forefront of his mind:

> *If there is one part of my public life on which I look back with more satisfaction than on anything else, it is with reference to the boon we conferred on third-class travellers. I have felt saddened to see third-class passengers shunted on to a siding in cold and bitter weather – a train containing amongst others many lightly-clad women and children – for the convenience of allowing the more comfortable and warmly-clad passengers to pass them. I have even known third-class trains to be shunted into a siding to allow express goods to pass.*
>
> *When the rich man travels, or if he lies in bed all day, his capital remains undiminished, and perhaps his income flows in all the same. But when the poor man travels, he has not only to pay his fare, but to sink his capital, for his time is his capital; and if he now consumes only five hours instead of ten in making a journey, he has saved five hours of time for useful labour – useful to himself, his family, and to society. And I think with even more pleasure of the comfort in travelling we have been able to confer on women and children. But it took twenty-five years to get it done.*

Not everybody appreciated the boon of cheap travel. An anonymous account, date unknown, tells how one man hitched a ride on the main-line train that ran between Euston and Liverpool. The man in question was apparently a sailor who chose to travel on the 9 p.m. express, for which the first stop was Rugby, some 82 miles up the line.

> *Mr Smith did not take his seat like an ordinary passenger inside any of the carriages but he travelled underneath one of them and would, no doubt, have concluded his journey to Liverpool in safety but that on the arrival of the train at Rugby the wheel-examiner, seeing a man's legs protruding from under one of the carriages, had the curiosity to make further search and discovered Mr Smith coiled round the brake-rod, a piece of iron not above three inches broad, in a fantastic position.*

Section of a London Underground station: illustration from *L'Universe illustré*, late 19th century.

Mr Smith was immediately uncoiled and being technically in error was detained in custody. The bottom of the carriage was only eighteen inches from the ground and where the engine takes up the water as it travels, Mr Smith was not more than six inches from the trough.

Magistrates fined him two shillings and sixpence, or 14 days in jail, expressing wonder at how he endured the ride. He presumably was able to assume the unorthodox position because Euston Station was thronged with people. And for London the transport problem would only get worse.

The answer as far as the Select Committee was concerned was to venture underground. Authorisation for this had been given in 1854 but the project was delayed by concerns about finance.

The world's first underground railway would link Paddington, the terminus of the Great Western Railway services, and Farringdon Street in the City in 1863. Along its four-mile route the existing road was lifted and a trench was dug and lined with bricks. Tracks were laid in the trench before it was enclosed once more and the road replaced. This elementary construction system was called 'cut and cover'. The ground-breaking first line was known as the Metropolitan or 'the Met'.

It was an immediate hit with Londoners, and some 30,000 people travelled on it daily. Nor was it an easy journey for those intrepid travellers. The wagons were initially open and drawn by steam engines. Although the route was amply vented it was nonetheless a smoky and dirty experience, albeit short-lived. The driver and guard, compelled to spend all day in the sulphurous atmosphere, were less fortunate still than the passengers.

UPRIGHT VICTORIANS WERE APPALLED THAT HEAVY DRINKERS AND ROUGH TALKERS WERE LABOURING OUTSIDE THEIR HOMES

In the following months further sections of underground lines were opened and extraordinary chaos was brought to London thoroughfares while construction work was underway. A phalanx of men turned up in the capital to undertake the work, and their reputation for hard living preceded them. Upright Victorians were appalled that heavy drinkers and rough talkers were labouring outside their homes, sharing the same streets.

But navvies – named for the so-called 'navigators' who built Britain's canals in the eighteenth century – were much maligned. If their appearance was ragged it was because the work they did was grindingly hard and dangerous. Many originated in Ireland and escaped to Britain at the time of the Famine. Often left at the margins of society, they developed a culture and even a language of their own. Hundreds had died as railways cut through the country, from both accident and disease. Construction companies frequently viewed the men as alarmingly expendable. Living quarters usually amounted to little more than a turf shack, and they were vulnerable to cold and hunger. Aside from gunpowder, the tools they had at their disposal to complete tremendous feats of engineering amounted to little more than picks and shovels.

Isambard Kingdom Brunel, who was known for being kindly rather than cruel to his workers, expressed no surprise when he was told 131 workers were taken to Bath hospital with serious injuries in less than two years while that part of the Great Western was under construction. 'I think it is a small list, considering the very heavy works and the immense amount of powder used,' he commented.

By the time the underground was being built, conditions had improved for navvies although it was still their muscle-power that turned the blueprint for the London underground into a reality. In 1868 the second underground railway opened, the District Line, and plans were afoot to link the two. The much-heralded Circle Line, which would finally provide a link between the major railway stations, was opened in its entirety by 1884.

Two years later, further up the Thames a short-lived cable-hauled train was opened between the Tower of London and Bermondsey. It was so unpopular that it was turned into a pedestrian subway until the opening of Tower Bridge in 1894, after which it was closed for lack of use. Today it still exists as a cable channel.

There was another step change in London's transport in 1890 when electric locomotives replaced steam. Tunnelling expertise was better than ever before too, and it led to the first deep tube running on what is now called the Northern Line. Its carriages were without windows as train operators reckoned there was nothing to see. Passengers relied on a guard to call out the station names.

Along London's Victoria Embankment, built from 1862 for traffic above ground and underground trains below, there's a pink granite monument from the heart of Egypt. Its purpose is to mark Nelson's victory over Napoleon at the Battle of the Nile in 1798. The battle was already a dim and distant memory, though, by the time the monolith was put up 80 years later.

Lying in the sand at Heliopolis, the 68-foot-tall monument, weighing 180 tons, was given to the British in 1819 by Mehemet Ali, the Albanian-born Egyptian leader who had himself fought against Napoleon. It was a somewhat belated gesture of thanks, a mere 21 years after the battle took place. Alas, he didn't come up with a suitable mode of transport. And, for the record, there's no evidence that the monolith, dating back to 1500 bc, has any links with the famous Egyptian queen Cleopatra, despite being commonly known as 'Cleopatra's Needle'.

Its transfer between countries didn't take place until 1877, when two bold Victorians grappled with the logistics. They were Sir William Wilson, who sponsored the operation, and engineer John Dixon, who designed a ship that would carry the stone from Egypt to Britain. Looking like an early container ship, its boxy appearance seemed

Engraving showing the obelisk ship *Cleopatra*, with Cleopatra's Needle aboard, off Westminster Bridge, London, 1878.

the ideal solution to the problem of protecting the monument in the heaving seas of the Bay of Biscay.

But on 14 October 1877 the operation looked doomed to failure. The ship, called *Cleopatra* in honour of its cargo, seemed to be sinking. Worse still, six men dispatched in a rowing boat from the tug towing it to rescue its crew were lost. Eventually, the crew of the *Cleopatra* were pulled to safety and the monument was abandoned to its fate. However, against the odds, it survived the storm and was later spotted by another ship which towed it to safety.

When the monument was finally put up by the River Thames it was guarded by two faux sphinxes and perched atop a time capsule containing, among other things, a portrait of Queen Victoria, pictures of a dozen beautiful women, a box of cigars, a hydraulic jack and a current copy of *Bradshaw's Monthly Railway Guide*.

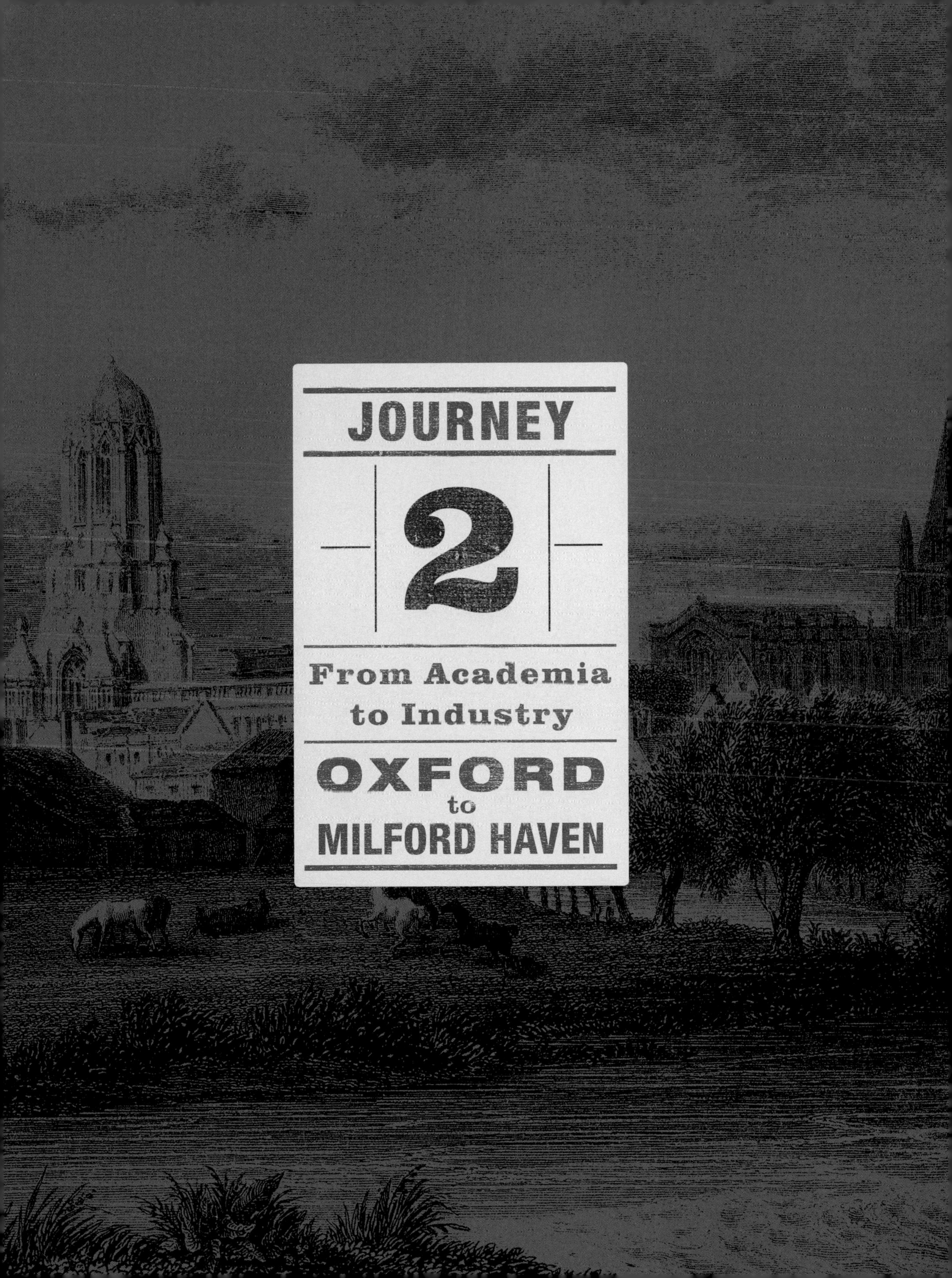
JOURNEY
2
From Academia
to Industry
OXFORD
to
MILFORD HAVEN

hen it comes to Oxford, Bradshaw's guide is fulsome in praise for its stylish architecture and noble seats of learning, all lying snug in a soft blanket of verdant countryside.

> **It is situated on a gentle eminence in a rich valley between the rivers Cherwell and Isis and is surrounded by highly cultivated scenery – the prospect being bounded by an amphitheatre of hills.**
>
> **From the neighbouring heights the city presents a very imposing appearance, from the number and variety of its spires, domes and public edifices while these structures, from their magnitude and splendid architecture, give it on a near approach an air of great magnificence...**
>
> **The high street in Oxford is justly considered the finest in England from its length and breadth, the number and elegance of its public buildings and its remarkable curvature which, from continually presenting new combinations of magnificent objects to the eye, produces an uncommonly striking effect.**

Perhaps all of this is to be expected from a city that has led education worldwide since the era of King Alfred. Yet, for all its elaborate grandeur, Oxford's reputation lay elsewhere for the majority of ordinary Victorians – for it was once as celebrated for its sausages as it was for its honours degrees.

Even *Bradshaw's* makes reference to the meatier side of Oxford's fame, almost in the same breath as speaking about its visible charms. 'Oxford has long been famous for good sausages and brawn.' If there's any doubt about the status of the city's sausages, then one only has to refer to Mrs Beeton's cookbook, one of the few publications as popular as *Bradshaw's*, to be certain.

Next to Queen Victoria, cookbook icon Mrs Isabella Beeton remains the most prominent woman of the nineteenth century. Mrs Beeton was the original domestic goddess — an orchestrator of recipes, an authority on cookery techniques and a fount of kitchen knowledge. Although she was dead by the age of 28 she left a mark

PREAVIOUS PAGE: A view of Oxford from meadows near the railway station, c. 1840.

Oxford High Street, c. 1890.

Local sausages on sale in Oxford's Covered Market, 1909.

that subsequent chefs – connoisseur, celebrity and quirky – have failed to erase.

In fact there were several remarkable facets to Isabella Beeton. She was, for example, the eldest of no fewer than 21 children. Her mother, Elizabeth, had four children before being widowed. Then she married widower Henry Dorling, who also had four children, and with him went on to have a further 13. That Isabella had plenty of hands-on experience in home skills as the oldest child of this immense brood is beyond doubt.

Mrs Beeton's areas of expertise extended beyond child-rearing and cookery. Her most famous tome, *The Book of Household Management*, included sensible advice on how to manage servants, medicines, poisons, animal husbandry and fashion. If that wasn't enough, she was also a talented pianist.

Perhaps strangely, she married a man who, like her, was born in Milk Street, in the City of London. Their mothers had previously been neighbours and friends. Crucially, Samuel Orchart Beeton was a publisher, and Isabella regularly wrote articles on housekeeping and cookery for him. When these were drawn together under one cover by her astute husband and published on Christmas Day 1861, a best-seller that endured for decades was created. Its full title was something of a mouthful, however: *The Book of Household Management Comprising Information for the Mistress, Housekeeper, Cook, Kitchen-maid, Butler, Footman, Coachman, Valet, Upper and under house-maids, Lady's-maid,*

Maid-of-all-work, Laundry-maid, Nurse and nurse-maid, Monthly, wet, and sick nurses, etc. etc. also, sanitary, medical, & legal memoranda; with a history of the origin, properties, and uses of all things connected with home life and comfort.

In it, perhaps for the first time, recipes were documented with exact quantities. From this sound foundation we can glean much about the Victorian diet. The book contains a recipe for Oxford sausages, the aroma of which so often floated around the city's dreamy spires. Mrs Beeton's recipe, which she claims as her own, is as follows:

> INGREDIENTS.— 1 lb. of pork, fat and lean, without skin or gristle; 1 lb. of lean veal, 1 lb. of beef suet, ½ lb. of bread crumbs, the rind of ½ lemon, 1 small nutmeg, 6 sage-leaves, 1 teaspoonful of pepper, 2 teaspoonfuls of salt, ½ teaspoonful of savory, ½ teaspoonful of marjoram.
>
> *Mode.*— Chop the pork, veal, and suet finely together, add the bread crumbs, lemon-peel (which should be well minced), and a small nutmeg grated. Wash and chop the sage-leaves very finely; add these with the remaining ingredients to the sausage-meat, and when thoroughly mixed, either put the meat into skins, or, when wanted for table, form it into little cakes, which should be floured and fried.
>
> *Average cost*, for this quantity, 2s. 6d.
>
> *Sufficient* for about 30 moderate-sized sausages.
>
> *Seasonable* from October to March.

ABOVE: **Isabella Mary Beeton (1836–1865).**

OVERLEAF: **The Shakespeare Express at Stratford-upon-Avon.**

There's been speculation that Mrs Beeton copied recipes rather than created them. Even if that's true, her plagiarism has given subsequent generations a unique view of the Victorian kitchen. Her common sense and forthright thoughts were her own, though:

> *Dining is the privilege of civilization. The rank which people occupy in the grand scale may be measured by their way of taking their meals, as well as by their way of treating women. The nation which knows how to dine has learnt the leading lesson of progress.*

Indeed, progress was the watchword of the Victorian era. The same industrial innovation that was powering trains was also present in the publishing industry. As steam-powered presses produced more books at cheaper-than-before prices, Mrs Beeton's words spread further than ever. In any event, after the railway reached Oxford in 1844 it is safe to assume that the city's sausages of distinction were no longer the preserve of scholars and dons.

Isabella Beeton died in 1865 after giving birth to her fourth child (the two eldest died as infants). Her husband survived her for a dozen years before succumbing to tuberculosis aged 46.

There are two more points of interest about Oxford in *Bradshaw's*. First, the book notes: 'The railways and the meadows around the city were all under water in the floods of 1853.' Secondly, it observes that in 1866 the city returned two MPs to Parliament and another two were elected by the university.

At first glance the fact that the university elite wielded such power seems like a scandal. In fact, voting was far from universal at the time *Bradshaw's* was published. After the much-heralded 1832 Reform Act it was still only one in seven men who were entitled to vote. Only with the 1867 Reform Act – passed a year after our *Bradshaw's* was published – was the vote extended to all male householders in urban areas. An amendment to it meant that more men than ever before could vote in elections after 1884, but that number still only represented six out of every ten men in the UK. Giving women the vote was a largely overlooked issue in Victorian England.

Heading north-west, the Oxford, Worcester & Wolverhampton line, finished by 1853, scythed through some of the country's prettiest scenes. By 1859 a branch line had been built off it from Honeybourne to Stratford-upon-Avon.

For Victorian sightseers a love of Shakespeare put Stratford high on the 'to do' list. *Bradshaw's* sheds considerable light on how the town fared 150 years ago:

> **This interesting part of Warwickshire is directly accessible by a branch of the Oxford, Worcester and Wolverhampton line by which means it is within about 100 miles journey by rail from London...**

THE SHAKESPEARE EXPRESS
4965

A stereoscopic postcard of Henley Street, Stratford-upon-Avon, Warwickshire, late 19th century.

> It is a municipal borough but derives its chief importance from being the birthplace of Shakespeare who was born here on 23 April (St George's Day) 1564 in an old-fashioned timbered house, opposite the Falcon, in Henley Street which, after some changes and the risk even of being transferred as it stood to America by a calculating speculation, was at last purchased by the Shakespeare Club and adopted by Government as a tribute to his memory.

Bradshaw's pays tribute to the luxuriant scenery surrounding Stratford but is quickly overcome with Bard fever.

> But rich and pleasant as the prospect is, it takes its crowning glory from the immortal poet, the mighty genius whose dust reposes at our feet. It is his genial spirit which pervades and sanctifies the scene and every spot on which the eye can rest claims some association with his life. We tread the very ground that he has trod a thousand times and feel as he has felt...

It also reveals some literary vandalism that occurred a century previously on the outskirts of Stratford:

The main entrance to the Lea & Perrins factory in Worcester.

> Ingon House was Shakespeare's at his death, having been bought by him in 1597 ... In 1752 it was sold to Mr Gathell, a Lichfield clergyman, who, on account of a dispute about the rating, pulled it down. He had already cut down the poet's tree, to save himself the trouble of showing it to visitors. Fortunately a cutting was planted by Garrick over the grave of Shakespeare's favourite grand-child Lady Barnard in Abington churchyard (near Northampton) and another is said to be at East Cliffe (Hastings), while the remains of the desecrated tree were sold in the shape of boxes, cups &c.

It's difficult to ascertain how much of the latter paragraph is true, but it remains a fascinating Victorian perspective on the well-worn story of Shakespeare.

A culinary theme continues further along the journey of the West Midland section of the Great Western Railway at Worcester. There's no indication that Mrs Beeton advocated the use of Worcestershire sauce, launched on unsuspecting British taste-buds in the 1830s. But it is the cornerstone of a British love affair with Indian food – albeit one with imperialist overtones – that has thrived ever since.

There is some confusion about the exact timeline of the sauce that is now a staple of the British dinner table, but there's consensus that

EDWARD ELGAR (1857–1934)

Worcester and its environs remained an inspiration to one of England's greatest composers. Edward Elgar was born 20 years after Victoria took the throne. His magnificent *Pomp and Circumstance* marches sound like a series of compositions from a man at the centre of British cultural life. With the first of the military marches also known as 'Land of Hope of Glory', his music is strongly associated with patriotism and imperialism.

Yet Elgar felt an outsider throughout his life and later deplored the use of his music to rally troops during the First World War, to fight for the British Establishment. He was a Catholic at a time when Britain was predominantly Protestant, although he had lost his faith by the time of his death. He lived in rural Worcestershire when the music world pivoted around London. And he was the son of a tradesman rather than a country gentleman, born in Lower Broadheath, Worcestershire, in a small house to a father who was a music dealer and piano tuner.

Elgar grew up with a hands-on mastery of music, quickly learning the piano, violin, cello, bassoon and trombone. Initial plans to work in a solicitor's office were abandoned in favour of a job as a music teacher. At 22 he became bandmaster at the local pauper's asylum, working for a superintendant who was convinced the sound of music would soothe some mental torment.

His output as a composer was at first limited and he only produced choral pieces and cantatas. He met his wife Caroline Alice when she became a pupil in piano accompaniment.

Her family were opposed to the relationship on the grounds of Elgar's class and faith. For her he wrote the popular 'Salut d'Amour' but sold the rights to it early on. Much later, he heard a busker play the tune outside the Queen's Hall in London and thought that the musician probably made more money from the piece than he ever had. When he asked the man if he knew its title, the violinist replied: 'Yes, it's "Sally, Damn her".'

His wife, who was nine years older than Elgar, devoted herself to his career, tidying his desk and maintaining a strict silence when he worked. For a while the couple lived in London, sure that Elgar would soon be fêted by the necessary musical impresarios. In fact the era was disastrously unproductive and they returned to the country, where Elgar found his greatest inspiration. It was spectacular scenery and gentle nature that stimulated his creativity. In

NOTES RETRIEVED AFTER HIS DEATH REVEAL THAT, FOR A JOURNEY TO LONDON, HE WOULD GET UP AT SIX AND WALK TO THE STATION AT MALVERN IN TIME FOR A SEVEN O'CLOCK TRAIN.

Sir Edward Elgar in his work room during his later years.

his remaining years he would take long walks and even longer cycle rides in search of musical revelation.

But Elgar knew he could not afford to sever links with the London music scene if his career as a composer was to flourish. As a result he used the train to visit London for day trips, keeping in touch with the musical trends that evolved there. The most important concerts took place at Crystal Palace.

Notes retrieved after his death reveal that, for a journey to London, he would get up at 6 and walk to the station at Malvern in time for a 7 o'clock train. By 11 a.m. he would have arrived at Paddington, and then he'd take the Underground to Crystal Palace via Victoria. Customarily, he listened to a rehearsal in the afternoon and a concert scheduled for tea-time so he could return home by the same route that evening.

In 1901 the first *Pomp and Circumstance* march was performed for the first time. It was, said Elgar, 'a tune that will knock 'em flat'. His prediction was spot on. Later, words with the title 'Land of Hope and Glory' were attached to the first march, although for Elgar the music became a wearying cliché with worryingly nationalistic overtones. Today *Pomp and Circumstance No. 1* is best known for its inevitable appearance at the London Proms and in graduation ceremonies across America.

Yet perhaps one of his most important contributions to today's musical world had yet to be made. After 1926 Elgar was able to record almost all his major works for the age of the gramophone, the first composer to do so. It provides a remarkable musical heritage, all the more valuable as Elgar's music had begun to go out of style even in his lifetime.

two chemists from Worcester, John Wheeley Lea and William Henry Perrins, concocted the first brew from a recipe brought back from India at the request of a local man who'd served there. The military type who provided it was brought in to sample the men's newly-made infusion – and spat it out immediately. It was, he declared, nothing like the spicy mix he had come to know and love in India.

Defeated by the endeavour, the chemists stowed the barrel containing the sauce in their cellar and only discovered it again much later. They decided to taste the sauce one more time before consigning it to a ditch, and found that it had matured into something altogether more palatable.

By 1837, the year Victoria came to the throne, Worcestershire sauce was being marketed, and soon it was popular worldwide. Locally it is better known as Worcester sauce. The recipe remained a closely guarded secret for years, but what is presumed to be original notes were discovered in a skip during the twentieth century and they throw partial light on it. In Victorian times Worcestershire sauce apparently contained cloves, salt, sugar, soy, fish, vinegar, essence of lemon, peppers, pickles and tamoraide – the Victorian spelling of tamarind.

Worcester wasn't famous for its sauce alone, however. Once the city was home to an army of glovemakers whose craftsmanship was celebrated worldwide. However, just as the sauce was coming into vogue, Worcester's glove trade was heading into a slow but certain decline.

British dominance in manufacturing faced the beginning of the end in 1826, when restrictions on foreign imports were lifted by the government as it embraced free trade. Initially, the law change wasn't disastrous. An estimated 30,000 people in Worcester were employed by 150 different glovemaking companies at the time, about half the glovemakers in Britain. But inevitably there followed death by a thousand cuts. The companies that survived had to modernise, and the outworkers in cottages spread around the Worcestershire countryside were usurped by factories.

In *Bradshaw's* there's a claim that, 40 years after the tariffs on imported gloves were lifted, Worcester still produced 'half a million pairs of leather and kid gloves annually, employing between one thousand and two thousand persons'. In fact, Fownes, one of the

RIGHT: 'The Royal Worcester Porcelain Works', c. 1880,from *Great Industries of Britain Vol. 1*, published by Cassell, Petter and Galpin. The making of Royal Worcester porcelain was another significant local industry.

OVERLEAF: A view of West Malvern.

country's leading glovemakers, didn't open its Worcester factory until 1887. But the line on the production chart was heading irrevocably downwards. Fownes survived until 1974, when the business transferred to Warminster in Wiltshire. But already Worcester's reputation for glove-making had long been consigned to history.

More than simply a fashion item, wearing gloves led to good health as they were a barrier to infection. Gloves also provided warmth in the days before efficient heating systems at home, in shops or on transport systems.

AMONG THOSE WHO EXPERIENCED THE WATER CURE IN MALVERN WERE FLORENCE NIGHTINGALE, CHARLES DICKENS AND CHARLES DARWIN

From Worcester, the line runs down to Malvern. Today's visitors to Malvern are usually keen to mimic Edward Elgar and stride out on the ancient granite hills which loom there. However, when Queen Victoria was a young woman the visitors were a different breed. Sickly, frail and optimistic that a newly fashioned water cure in Malvern would set them on the road to recovery, they arrived by carriage. Perhaps curiously, Malvern arrived late to the railway party despite its enviable reputation.

It is quite likely contaminated water may have been the source of numerous ailments at the time. London's water was notoriously noxious and other centres of urban population were distinguished by insanitary conditions.

In Malvern the spring water flowed clean and pure. That was not lost on one W. Addison, surgeon to the Duchess of Kent, Queen Victoria's mother. In 1828 the *Quarterly Journal of Science, Literature and Art* produced by the Royal Institution of Great Britain gave a warm reception to Addison's investigation into the health-giving qualities of Malvern water.

> *Mr Addison's work is scientific and ingenious; he attributes the many extraordinary recoveries which have occurred at Malvern partly to the salubrity of the air and partly to the purity of the water which from the analysis he has given of it seems to contain much less saline and earthy matter than any we are acquainted with and we think he has laboured with considerable success to prove that the continued use of a pure water may be a powerful means of removing or preventing many chronic disorders.*

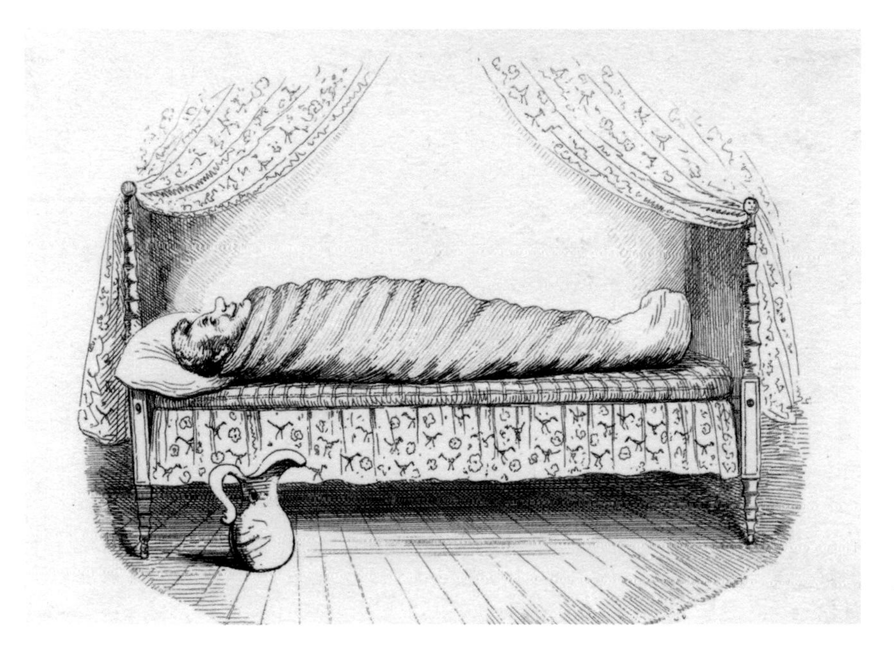

The Malvern Water Cure.

With water cures something of a fashion in Europe, Malvern was soon on the radar of two doctors who had enduring faith in the hidden powers emanating from the filtered rainwater that bubbled up in the form of springs. Dr James Wilson and Dr James Gully were established in the town by 1842, a full 18 years before the railway arrived in Malvern, rigorously ensuring their patients exercised, ate properly, drank spring water and, more controversially, were douched in hot and cold water.

The regime was largely an improvement on prevailing lifestyles and the results garnered by the doctors were encouraging. Among those who experienced the water cure in Malvern were Florence Nightingale, Charles Dickens and Charles Darwin. To get there, they used carriages.

Bradshaw's guide speaks well of the cures:

> **St Anne's and Holywell, springs much resorted to, are slightly tepid and sulphurated and useful, especially in glandular and skin complaints.**

But it reserves its greatest praise for the surrounding scenery.

> **The prospect from the hills embraces part of eight or nine counties, including the vales of the Severn and Evesham, or the Avon, the cathedrals of Worcester, Gloucester and Hereford, Tewkesbury Minster and the Welsh Hills &c and is the finest in the kingdom.**

Water bottled in the Malvern Hills was taken to London's Great Exhibition in 1851 and presented to Queen Victoria. It provided the groundwork for the bottled water industry which is today worth about £1.4 billion in Britain.

In time the railway reached this corner of west England. The Worcester & Hereford Railway Act was passed in 1853 and, seven years later, the line from Worcester to Malvern was open for business. A year after that the railway tracks were laid beneath the Malvern Hills, extending the line to Hereford.

Today the station at Malvern is remarkable for its Victorian embellishments. It was designed by a local architect, Edmund Wallace Elmslie, apparently to the specifications of Lady Emily Foley, the lady of the Malvern manor.

Lady Foley was the daughter of a duke and for 54 years the widow of MP Edward Foley. Even by Victorian standards she was considered

Great Malvern Station, Worcestershire, England.

a harridan. It is said that when she dined alone at her home in Stoke Edith she was waited on by a butler and four footmen. When she visited tenant farmers and their families she expected a bow from the men and a low curtsey from the women she met. Habitually, no one entered church for Sunday services until she had taken her seat. Notwithstanding, her influence on the architecture in Malvern, both in the town and at the station – which she wanted to be better than any other except for the terminus – has meant a legacy of grandeur. Lady Emily died as the twentieth century dawned, aged 94.

Lady Emily Foley, c. 1840.

Going on from Malvern and Hereford the line swings across the border into Wales, where a rich green environment is soon eaten up by industry. Beyond Hereford on the West Midland main line is Abergavenny, about which *Bradshaw's* shares some compelling history. The guidebook reveals:

> **Abergavenny became celebrated for its Welsh wigs, made of goats' hair, some of which sold at 40 guineas each. Physicians also used to send patients here to drink goats' whey.**

Although it isn't specific about when the goat ruled in Abergavenny, the guide is presumably referring to Regency times when wigs were fashionable for men and women. As for Victorian Abergavenny, its outlook was much more industrial.

> **Its present prosperity arises from its flannel weaving and the valuable coal and iron works at Clydach, Blaenavon &c in its neighbourhood – a state of things likely to be much increased by the Newport, Abergavenny and Hereford Railway, part of that important chain which unites South Wales to Liverpool and the North of England.**

At the time this entry was written, the Newport, Abergavenny & Hereford Railway had been open for 13 years, having been sponsored by London & North Western Railway. In 1860 it merged with the

Oxford, Worcester & Wolverhampton Railway and the Worcester & Hereford Railway to form the West Midland Railway.

On the English side of the border, in the Forest of Dean, ancient rights laid down by law have for centuries given local men the right to begin mining what was once a rich coal seam. In effect, anyone born in the Forest who had been working as a miner for a year could start their own business. It was also their right to sell their mines to whomever they chose. Even by Victorian times many of these small mines had passed into the hands of bigger companies. Today, with the heyday of coal-mining well and truly at an end, only a redoubtable few continue. But evidence of the unique way of life that once existed there remains in the form of the tram rails built by miners for their own wagons.

By contrast, in nearby South Wales coal mines were proliferating. Alas for the local population, these were privately owned by people or companies that cared more for profit margins than for the men on their payrolls. Most of the 500 or more mines that peppered the landscape of South, Mid and West Wales were started during Victorian times, when coal was the principal fuel of steam engines on the railways, on ships and in factories. Life in the mines was at best unpleasant and at worst exceedingly dangerous.

At the Ferndale Colliery in the Rhondda, for example, 178 people died – including at least three aged 13 – in an accident on 8 November 1867. It happened just months after the opening of the pit. For weeks the bodies of the dead were being brought to the surface in a slow and painful recovery operation, most burnt beyond recognition. The two explosions that occurred were thought to have been caused by an accumulation of gas. Nineteen months later another 53 died in the same pit and the blast which killed them was also fuelled by gas.

Thirteen miles north of Cardiff in the village of Cilfynydd there were 290 deaths caused by an underground explosion on 23 June 1894, the most ever killed in a mine at the time. The youngest of the dead was 14; only 24 of the victims were aged 40 or above, and there were 25 who shared the surname of Jones. In a small, interwoven community, not a household was spared the grief of sudden and brutal loss. Of the 125 horses working underground at the time, only two survived. Nor were these isolated incidents. Between 1851 and 1920 there were 48 disasters in the South Wales coalfield, and 3,000 deaths.

AT THE FERNDALE COLLIERY IN THE RHONDDA, 178 PEOPLE DIED – INCLUDING AT LEAST THREE AGED 13 – IN AN ACCIDENT ON 8 NOVEMBER 1867

A view of Ferndale Colliery (also known as Blaenllechau Colliery) and Tylorstown, Rhondda Valley, Glamorgan.

The life of a Welsh mining village during Victorian times was captured in the novel *How Green Was My Valley* by Richard Llewellyn, published in 1939. The author based the story on conversations he had with miners from the era.

Some advantages of the steam engine were brought to bear in coal-mining, in terms of pumping out water and winding up loads of coal. But largely working conditions were primitive and those that laboured in them included men, women and children.

In 1842 the Royal Commission report on conditions for women and children in mines caused concern in polite society, not least for its graphic illustrations. There was, the Commission said, 'cruel slaving revolting to humanity'. While eight-year-olds were commonly employed, it said, the starting age for mineworkers was often as low as five. There swiftly followed the Mines and Collieries Bill, prohibiting women, girls and boys under the age of 10 from working underground.

The Coal Mines Inspection Act of 1850 was an attempt to curtail the number of accidents, with inspectors appointed to check on the safety procedures in British mines. A decade later another law was passed, this time to enhance rules about safety and raise the age limit for boys working in mines from 10 to 12.

Doubtless the paperwork brought about some good effects, but lives were still lost in mining accidents every year, with many more men being injured underground. From 1872 Parliament insisted that pit managers had certificates to prove they were adequately trained in safety measures. Miners were also able to appoint inspectors among themselves, and the Mines Regulation Act of 1881 gave the go-ahead for parliamentary inquiries after accidents. Still, little could be done without the collaboration of pit owners.

No matter what the dangers, the coal industry snowballed in size and assumed extraordinary importance, as the appetite for coal appeared insatiable. What South Wales needed most was a transport link to a viable dockyard.

Cardiff seemed the obvious choice for such a dockyard. Given its size and significance today it is hard to imagine that the capital of Wales was, even at the end of the eighteenth century, just a humble shanty town with few prospects. No coal was exported from Cardiff as, everyone agreed, the operation would be too expensive to countenance.

In 1791 the Glamorganshire canal was built to bring iron ore from up-country mines, especially the Dowlais Ironworks at Merthyr Tydfil, to the coast at Cardiff. A sea lock was opened on the River Taff five years later. Yet still Cardiff was literally a backwater. The 1801 census numbers its population at just 1,870, making it only the 25th largest town in Wales at that time.

However, the world was on the cusp of great change as the age of the train began to unfold. For railways to succeed two commodities were needed in quantity: iron for rails and coal for fuel. And both were found in abundance in South Wales, which became a power house of the industrial revolution.

FOR RAILWAYS TO SUCCEED TWO COMMODITIES WERE NEEDED IN QUANTITY: IRON FOR RAILS AND COAL FOR FUEL

One man set about changing the face of Cardiff to greet a new industrial dawn. John Crichton-Stuart, 2nd Marquis of Bute and the premier entrepreneur of the age, sensed the potential of a port, and in 1839 Bute West Dock, the fruit of his investment, opened. For his foresight he has been dubbed the creator of modern Cardiff. But still the city lacked some of the essentials of practical industry.

The missing connection was, of course, transport between the burgeoning coal and iron ore mines and the newly functioning outlet for exports at Cardiff. The Glamorganshire canal was choked with traffic ferrying some 100,000 tons of coal a year into Cardiff, a third of which was bound for London by ship. But the national thirst for coal was far from sated. The breach was filled by the Taff Vale Railway (TVR), proposed in 1836 by the ironworks' owners and built by Isambard Kingdom Brunel.

In a departure from his customary style, Brunel decided against broad gauge, believing the wagons transporting cargo like coal and minerals could never safely travel as fast as broad gauge allowed.

Coal ships tied up at Cardiff Docks.

ABOVE: **John Calvert hosts a banquet for his workmen and their families from the Rhondda branch of the Taff Vale Railway, near Newbridge, Glamorgan, 14 August 1851.**

The terrain was also hilly and thus more suited to narrower gauges. The estimated cost of the TVR was not quite £191,000. With the proposed route mostly shadowing a river valley there were no mighty obstacles in its path but for one steep incline. At the beginning that section was operated by stationary engines hauling the wagons by cable, until a diversion made in 1864 allowed locomotives along the entire route. In essence, loaded wagons from the coalfields headed downhill to the docks.

The first stretch of the line between Cardiff and Abercynon opened in October 1840 and the iron road reached Merthyr the following year. In 1841 the first of numerous branch lines opened, linking the main TVR at Pontypridd – then known as Newbridge – with Dinas. Pontypridd became the beating heart of the line, with trains often rumbling through its station at a rate of two a minute and a platform that at one time was the longest in Britain. Its signal box was decked with no fewer than 230 levers.

When the TVR was completed in 1841 it became possible to bring loaded wagons from Merthyr Tydfil to Cardiff within an hour. Now coal from Wales was sold on domestic and overseas markets, making the area's future seem assured. The number of daily dedicated passenger services was raised from two to three in 1844, and the following year prices for tickets in each of the three available classes were dropped by a shilling.

BY THE 1851 CENSUS, WALES HAD BECOME THE FIRST COUNTRY REGISTERING MORE PEOPLE EMPLOYED IN INDUSTRY THAN IN AGRICULTURE

By the time of the 1851 census, Wales had become the first country registering more people employed in industry than in agriculture. But not everything was rosy among the railway operators, dock and mine owners. The TVR board felt the docks were slow to develop and sought a new outlet at Penarth. Meanwhile, the dock owners began backing other railway companies, although the TVR with its 124 miles of track remained the largest in the region (at its peak in 1897, it was carrying some 14 million tons of coal and coke, with its 216 locomotives covering some 2,800,000 miles a year). However, the TVR was limited in its scope, and not least by the short-sighted approach of its investors. Possessing only a single track, it soon became clogged with traffic much as the Glamorganshire canal had been before its arrival (the track was eventually doubled in 1857). The dock at Cardiff – which was incapacitated at times by the tide – was also often congested with cargo awaiting departure, even after the East Bute Dock was added in 1855.

Bradshaw's obviously believed the East Dock was an answer to the capacity crisis.

> The new Bute Docks made on a tract of wasteland by the Marquis of Bute who is lord of the manor are about one mile below the town, deep enough for ships, with a basin of one and a half acres and an entrance 45 ft wide. A ship canal 1,400 yards long, 67 yards wide, runs up to the town. The coal and iron of Merthyr Tydfil and the neighbourhood are the chief exports and the quantity almost doubles itself every two or three years.

Yet still, the amount of dock space was found to be lacking. In 1865 Penarth docks were built. The TVR leased the Penarth Harbour,

Dock & Railway and the Penarth Extension Railway, operational from 1878, but capacity remained an issue.

Meanwhile, money was continually pumped into the Cardiff dock, with Roath basin opening in 1874, followed 13 years later by Roath Dock. Compelled to modernise to keep pace with demand, the dividends paid to investors in Cardiff's docks were marginal despite what were perceived to be high dock charges levied upon users.

Mine owners in the Rhondda saw opportunities for expansion being squandered by the limitations of the infrastructure so recently built. Raising their eyes to the horizon, their gaze fell upon a location beyond Cardiff which they thought might solve their problems.

Barry Island was so small as to be inconsequential. In 1841 the census recorded just 104 residents. Yet it was ideally placed by the sea to offer up new trade routes, breaking the monopoly enjoyed thus far by Cardiff. Despite heated objections from the high-powered group behind Cardiff's railway and docks, assent for Barry's rival scheme was given in 1884 at the second time of asking.

Members of the Barry Railway Company used joined-up thinking to open docks and dedicated track in 1889. Alongside its ship-filled dock was a repair yard, warehousing, cold stores, a flour mill and an ice factory. Almost immediately, vast quantities of coal once destined for Cardiff or Penarth were diverted into Barry. The financial impact on the TVR and its shareholders was colossal, with dividends reduced from a booming 15 per cent to 3 per cent in just two years.

The TVR responded by building another line that would connect Cardiff, Penarth and Barry along the coast. For its part the Barry Railway Company constructed a further direct line between Barry and Penarth. Other companies were keen for a slice of the industrial action, too. Coming late to the table were the Rhondda & Swansea Bay Railway, incorporated in 1882, and the Pontypridd, Caerphilly & Newport Railway, from 1884. As a result, that corner of South Wales became etched with railway lines in the same way as the north east of England.

On the Barry Island Heritage line.

In 1891 Barry's population had reached 765 and was rising fast. A decade later the figure stood at 27,000 (and within another 20 years the population figure for Barry, which by now spilled over into near

neighbourhoods, was 40,000). By the time of Victoria's death Barry was poised to overtake Cardiff as the largest docks operator in the region.

At the same time Barry Island laid down the foundations of its tourist trade. It became a favourite destination not only for coal but for the miners who hacked it out, easily accessible by railways which were extended in 1896 to take visitors beyond the docks to the island itself.

It was fortunate this alternative thread in the economy was developed, as the heady days of Barry Docks were already counting down. By the end of the First World War the British economy took on a different hue, not least because Germany paid reparations to Britain in the form of coal. Less was therefore needed from the Welsh coal mines.

One of the trades most severely affected by this was that of the coal trimmers employed at both Barry and Cardiff. It was their job to ensure a cargo of coal was evenly spread in a ship's hold. A load that shifted suddenly at sea could destabilise a ship, with disastrous consequences. The Cardiff, Barry and Penarth Coal Trimmers' Union was formed in 1888 to strike a fair deal for men who worked below deck wielding shovels in a grim and dangerous environment. Coal trimmers were eventually made obsolete by larger, self-trimming ships and the collapse of the coal industry.

However, other unions met with more lasting success despite a grim struggle at the outset. One such union was the Amalgamated Society of Railway Servants (ASRS), founded in Birmingham in August 1871, the same year the Trade Union Act gave groups like this one some status and protection in law.

With the hazards of railway work claiming hundreds of lives each year, there was a desire among the men for safer working conditions. Some railway companies like the Great Western Railway had a reputation for being paternalistic and provided outings, schooling and religious meetings for their staff. But care for the well-being of its workers stopped short of tolerating union activity. After guards working for the Great Western Railway formed a Railway Working Men's Provident Association in 1865, its leaders were sacked.

Profits were still the main motive for railway companies, which often compromised basic safety to save money. One parliamentary inquiry of the era discovered many men were exhausted, working 90 to 100 hours each week. Consequently, most employees felt themselves overworked and at risk.

WITH THE HAZARDS OF RAILWAY WORK CLAIMING HUNDREDS OF LIVES EACH YEAR, THERE WAS A DESIRE AMONG THE MEN FOR SAFER WORKING CONDITIONS

ABOVE: **A steam train crosses the viaduct to Barry Island.**

OVERLEAF. **Workers at Cardiff Docks during the British Coal Strike.**

At the ASRS Charles Bassett-Vincent began organising railway workers, initially on an informal basis. He aired grievances by writing to newspapers and contacting the Trades Council. His aim was not to call for strikes – considered dangerously disloyal – but instead to enter into talks that would iron out disputes. (Bassett-Vincent went on to found the National Association of General Railway Clerks in 1897 and was its first general secretary.)

The new group was popular, given there was no real precedent for working men to follow at the time. Within a year some 17,000 men put their names to its register – only a small portion of the 250,000 who worked on railways nationwide, but it was a start. However, predictably, the new organisation was not only arguing with railway management but soon its members were arguing among themselves. Although it was aimed at everyone who worked in the railways there were sensitive demarcation lines recognised by certain sectors. Drivers and firemen, for example, refused to be grouped with other workers and formed their own Associated Society of Locomotive Engineers and Firemen in 1880.

By 1882 ASRS membership was pegged at 6,300, but the word continued to spread. The first meeting of the Merthyr branch of the

ASRS was held on 15 July 1888. Richard Evans was elected chairman and the society decided on monthly meetings. Among others, the Merthyr branch represented workers on the Rhymney and Taff Vale railways.

The initial reluctance to strike was gradually eroded in the face of intransigence among railway company management. And one such action by the men of the TVR led to a major court case that shunted the British Labour Party to the political foreground.

In 1900 signalmen on the TVR went on strike, followed by the enginemen. The ASRS made the strike official, believing it was protected by the 1871 Act. Infuriated, the TVR, driven by its general manager Ammon Beasley, took out a legal injunction to stop the strike and sued the ASRS, claiming huge losses. At the end of lengthy proceedings the union was ordered to pay £23,000 in damages and £19,000 costs after a hearing in the House of Lords held in 1901.

The Lords' ruling rendered unions everywhere unable to take action against employers, no matter how legitimate the cause. Strikes were probably going to mean bankruptcy. Working men now looked for a new brand of representation that would give them a voice. The ASRS had been one of the founders of the Labour Representation Committee in 1900, which was renamed the British Labour Party six years later. In the meantime the newly elected Liberal Government introduced the Trades Disputes Act in 1906, which gave unions the right to call strikes without the threat of legal action being used against them.

At the heart of Welsh industry was Merthyr Tydfil, loud and proud. According to *Bradshaw's*:

> **It stands up the Taff, among the ragged and barren-looking hills in the north-east corner of Glamorganshire, the richest county in Wales for mineral wealth.**
>
> **About a century ago the first iron works were established here, since which the extension has been amazingly rapid. Blast furnaces, forges and iron mills are scattered on all sides. Each iron furnace is about 55 ft high, containing 5,000 cubic feet; and capable of smelting 100 tons of pig iron**

L STRIKE - IDLE DOCK LABORERS, CARDIFF

weekly, and as there are upwards of 50, the annual quantity of metal may be tolerably estimated, but as great as that supply may seem it is scarcely equal to the demand created for it by railways.

Then there is an uncommon dose of heartfelt concern pouring forth from *Bradshaw's*.

Visitors should see the furnaces at night when the red glare of the flames produces an uncommonly striking effect.

Indeed the town is best seen at that time for by day it will be found dirty and irregularly built, without order or management, decent roads or footpaths, no supply of water and no public building of the least note, except barracks and a vast poor house lately finished in the shape of a cross on heaps of the rubbish accumulated from the pits and works. Cholera and fever are, of course, at home here in a scene which would shock even the most 'eminent defender of the filth'.

The rapid expansion of the town had not been accompanied by education, as the guidebook notes:

Out of 695 couples married in 1845 1,016 persons signed with marks...

We do hope that proper measures will be taken henceforth by those who draw enormous wealth from working these works to improve the condition of the people.

The guide implores ironworks owner Lady Charlotte Guest to become one of the 'Nightingale Sisterhood' in order to improve social conditions.

However, according to an obituary of Sir John Josiah Guest, Lady Guest's husband and the previous owner of Dowlais Ironworks, the couple had carried out ample amounts of charitable work. Minutes of the Institution of Civil Engineers after Guest's death in 1852 claim:

Although strict in enforcing subordination among the multitude of men in his employment, he was ever watchful for their interests, and sought their spiritual and temporal benefit in every way; founding places of worship, and establishing schools, whilst, during periods of mercantile depression and the visitation of disease, his charity was unbounded, and in these labours of love he was ably seconded by his wife, the Lady Charlotte Elizabeth Bertie, only sister of the Earl of Lindsey.

This estimable lady, whose literary powers are as well appreciated as her general talents and acquirements in branches of knowledge not usually presenting attractive features for ladies, appears to have felt, that on arriving among a dense population of hitherto ill-educated people, speaking a peculiar dialect, and isolated by their habits and manners, few of the inhabitants of Dowlais having ever travelled twenty miles from their homes, she was called to a task of no little

Factory smoke at Dowlais, South Wales.

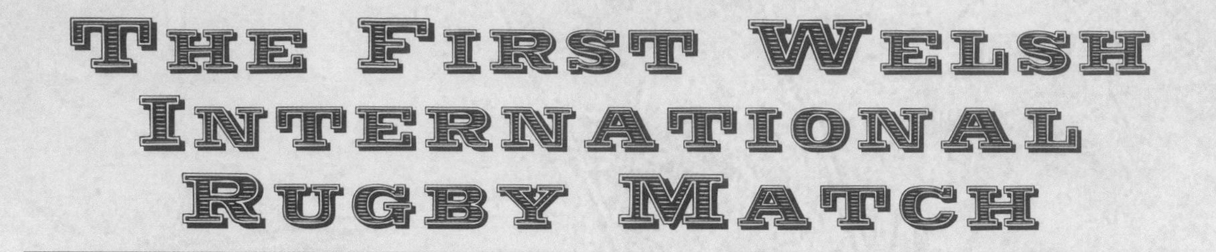

The First Welsh International Rugby Match

Men who laboured hard during the week in heavy industry, living in deprived circumstances, increasingly sought recreation at the weekends. Rugby was poised to make its journey from being a public-school sport to a working-man's pastime.

The Welsh Rugby Union was formed after the nation was humbled in its first international, as sports historian Patrick Casey explains:

> *The first international between England and Wales took place on 19 February 1881 at Blackheath.*
>
> *A game more noted for the chaotic organization of the Welsh side than anything else, it was Wales' first international. The players had never played together before.*
>
> *No formal invitations to play were sent out to the Welsh XV. Two of those expected to appear didn't turn up so bystanders, University undergraduates with tenuous Welsh links but who had travelled to London to see the match, had to be roped in to play for Wales.*
>
> *It also didn't help that the changing rooms were a local pub (The Princess of Wales, which remains to this day). Both teams had to walk the half a mile across the common to play. Rumour has it that the Welsh team needed some Dutch courage before the match so had been drinking heavily.*
>
> *The game was a farce. The Welsh were hopelessly outplayed and under modern scoring values lost 82–0.*
>
> *Richard Summers played in that match and said of the Welsh outfit: 'We played in ordinary, light walking boots with a bar of leather across the sole to help us swerve. Jerseys were fitted high at the neck with serge blue knickers fastened below the knee with four or five buttons. We changed at the Princess of Wales public house nearby.'*

Richard Mullock is credited by many as the man behind a meeting that established the Welsh Rugby Football Union held soon afterwards. It became the Welsh Rugby Union in 1934.

'Collared', an engraving from Shearman's book on rugby, c. 1887.

In a later game at Blackheath, England defeated Wales by only 17–0.

Newport-born Mullock was a keen sports fan. A father of six and a printer by trade, he was left to organise the side that faced England at Blackheath. His task was made much harder because Swansea were playing Llanelli the same day in a crucial and popular cup match.

Undeterred by the crushing defeat, he orchestrated a meeting with the eminent South Wales rugby clubs. The Welsh Rugby Football Union was formed on 12 March 1881 at the Castle Hotel, Neath, down the tracks from Merthyr, with Mullock elected honorary secretary. The new committee's first job was to select a team for the match against Ireland a month later.

Unfortunately, Mullock's later career was tarnished with accusations of financial impropriety made in 1892, although he had undoubtedly covered some of the organisation's expenses himself. Patrick Casey takes up the story again:

So he was forced to retire with Welsh rugby on the verge of international success, about to win its first Triple Crown.

In April 1893 he was declared a bankrupt but continued the family printing business. In 1902 Mullock was sued in court for £3 of goods. He pleaded that the £1 a month he earned from selling drawings was insufficient to support his wife and their six children. He left the Newport area and disappeared from public view. Rumour had it he had emigrated that year to Africa.

Richard Mullock died in St Thomas' Hospital, London, on 8 July 1920 after an operation for throat cancer. By then he was living and working as a printer's clerk in Chiswick.

importance, to which she lent all the powers of her mind: she acquired the language, she visited the homes, she administered to the wants of all around her; and in thus performing her Christian duties, she extended the previously-acquired influence and power of her husband.

A Bessemer Converter – a process designed for the mass production of steel from molten pig iron.

It's safe to assume that, even if she had carried out good works, there was still plenty more to do. Many of the workers at Merthyr had migrated from rural areas in Wales and were Welsh speakers. In common with other densely populated industrial towns in South Wales, housing was poor, sanitation sparing and disease rife. Some children received an education at the expense of local industrialists like Guest. When this occurred it was in English, making the region bilingual by the twentieth century.

The Dowlais Ironworks, founded in 1759, was the biggest in the world during Victorian times. Thanks to Dowlais and other ironworks, Merthyr was the biggest town in Wales by 1801. Forty years later it had 18 furnaces, 9,000 workers and 100 steam engines. When the King of Prussia visited in 1844 he called Merthyr 'the fiery city of Pluto'.

Henry Bessemer's 1855 'Converter' process to turn pig iron into steel secured the fortunes of the Dowlais Ironworks. Steel could be produced cheaply and in quantity, and by 1890 Britain was the largest producer of steel in the world, with South Wales at the core of production. The Converter only worked where phosphorus-free iron was used, and that was in plentiful supply in South Wales; industrial centres in Europe and America struggled to take advantage of the process because their mineral wealth contained phosphorus.

However, Britain's iron rule was short-lived, thanks to advances made by two cousins in Blaenavon, a short rail trip away from Merthyr.

Percy Carlyle Gilchrist, who was the works chemist at the Blaenavon ironworks, and his cousin Sidney Gilchrist Thomas, studied the Bessemer Converter – which had changed Blaenavon's struggling fortunes. They discovered that a lining of special bricks made of limestone and fireproof tar enabled the Converter to change iron that contained phosphorus into malleable steel.

From these results, publicised in 1878, the pair made a fortune. But inadvertently they spelled the end of Britain's domination of steel production. Now other towns on top of iron ore deposits could get to work producing steel, including those in America

and Germany. And by 1902 Britain was ranked third among steel-producing countries for production, and saw its export market diminishing by the year.

A vintage portrait engraving print of Sir William Hamilton by Henry Hudson, c. 1788.

St David's was left off the railway map, with Haverfordwest being the nearest station. According to *Bradshaw's*:

> **...the road between Haverfordwest and St David's is the most execrable in the United Kingdom but replete with scenery magnificently grand.**

Roads had become something of a sore point in West Wales by Victorian times. In fact, it was not so much the roads but the tolls local people were forced to pay to use them. Farmers were scratching a living from a landscape that wasn't always forgiving. They needed lime to improve the soil and a ready access to market to sell their produce. Turnpike Trusts, established by Acts of Parliament, were intended to build roads and charge a small fee to users for their maintenance, but typically the roads were poorly maintained and the tolls were high. There were 11 Turnpike Trusts around Carmarthen alone, and sometimes farmers encountered several turnpikes on the same road, reducing their income in times of high taxation.

After a new turnpike was introduced on a road that linked farmers from West Wales to their local lime kiln in 1839, there was an uprising of an unusual kind. Men dressed as women, with blackened faces, turned up to smash the turnpike hut and began what became known as the 'Rebecca riots'.

No one knows for sure why these men assumed the identity of women. There was a long-standing custom in the area to put on women's clothes and darken faces for public parades. It was also symbolic of the world being upside-down. Some historians have linked it to a line in the Old Testament in which Rebecca is told, 'Let thy seed possess the gates of those which hate them'. A more down-to-earth explanation is that the leader of the first riot borrowed his garb from a woman called Rebecca.

In any event, the Rebecca riots escalated, and in 1842 a letter purporting to be from those involved warned: 'As for the constable and

A portrait of Lady Hamilton.

the policemen, Becca and her children pay no more attention to them than the grasshoppers which fly in the summer.'

On 26 May 1843 rioters went into Carmarthen to attack the workhouse – ostensibly built to give work to the poor but in reality akin to a jail. By now the authorities were gravely concerned. In July that year an officer insisted, 'Great numbers of discharged workmen from Merthyr and Dowlais have come into the county and are active in persuading the people to mischief'.

From folk hero status, Rebecca – who might well have been more than one man – and his colleagues began to spread fear in their communities as they took the law into their own hands. The death of an elderly woman turnpike-keeper turned the tide of public opinion against the rioters.

It was their good fortune that a reporter for *The Times*, T. C. Foster, had shone a light on their grievances. Shortly afterwards a Commission of Enquiry halved the tolls and the bizarre public protesting faded into history.

At the end of the line is Milford Haven. The arrival of the railway here in 1863 gave new life to a run-down town. At the town's inception some 70 years before it had been touched by celebrity and scandal. Its founder was the diplomat Sir William Hamilton, who was utilising land bequeathed to him on the death of his first wife. In 1791 he married for a second time, aged 60. His wife was 26-year-old blacksmith's daughter Emma Hart, a talented dancer and noted beauty. She was later considerably more famous as Lady Emma Hamilton, the mistress of British naval hero Lord Nelson. Nelson travelled with the Hamiltons, his presence tolerated and perhaps even encouraged by Sir William. They even visited Milford Haven together in 1802, when it was newly established as a whaling port with crews drawn from Nantucket. There were public celebrations to mark the occasion. Obligingly the Navy Board also established a dock there.

However, in 1810 the whalers returned to America, the same year Milford's bank collapsed. Four years later the Navy relocated to Pembroke and Milford was left in the doldrums. In 1874 Milford docks were under construction, laying the groundwork for the fishing industry which would dominate for decades. Only after the development of giant tankers in the 1950s did Milford become the focus of the oil transport industry.

JOURNEY 3

A Royal Progress

WINDSOR to WEYMOUTH

Initially Queen Victoria was not a fan of railways. Although she was a child when the first railways began to operate, she still subscribed to the commonly held view that people could disintegrate and die under the pressure of excessive speed. However, her more scientifically minded husband, Prince Albert, had a more far-sighted view of the invention, although he was evidently a cautious passenger. He memorably uttered after one leisurely trip: 'Not quite so fast next time, Mr Conductor, if you please.'

After a delay brought about by concerns the railway would blight the castle, two stations were built in short order in Windsor by rival companies hoping to curry royal favour. Great Western Railways was laying tracks into Windsor from Slough while South Western Railway was heading in from neighbouring Datchet.

Both projects were running neck and neck until iron supports on a bridge between Windsor and Datchet wobbled, permitting Great Western Railway with its rock-solid Brunel bridge across the Thames to win the railway race. Windsor & Eton Central Station opened on 8 October 1849 and Queen Victoria made her first journey from it on 23 November. Within four weeks Windsor & Eton Riverside Station was in business, although its opening was a somewhat more downbeat occasion.

The fact that railways reached Windsor at all was a cause for celebration, as the authorities at nearby Eton College, founded by King Henry VI in 1440, were implacably opposed to any trains running in the vicinity, fearing they would be a distraction for students. At the time the college was an immensely powerful institution with plenty of friends in Parliament, where plans for new railways were agreed before they were built.

However, concerns that trains would provide a ready escape route to London for listless students, or become a target for stone-throwing scholars, proved unfounded. In fact, two Eton schoolboys became royal heroes one day after the Queen returned from a train trip.

On 2 March 1882 a disgruntled would-be poet, Roderick Maclean, fired a pistol at Victoria as her carriage was leaving Windsor railway station. Earlier on in her reign the Queen was not universally popular, after withdrawing from public life following the death of her husband. But by now she had been wooed back by politician Benjamin Disraeli and was more highly regarded than ever. Maclean's gripe was that

PREVIOUS PAGE: **Queen Victoria and Prince Albert welcome Napoleon III and Empress Eugenie to Windsor Castle.**

RIGHT: **Roderick Maclean's attempt to assassinate Queen Victoria at Windsor Station.**

OVERLEAF: **Windsor Station, 1908.**

THE ILLUSTRATED LONDON NEWS

REGISTERED AT THE GENERAL POST-OFFICE FOR TRANSMISSION ABROAD.

No. 2236.—VOL. LXXX. SATURDAY, MARCH 11, 1882. WITH TWO SUPPLEMENTS SIXPENCE. BY POST, 6½D.

she had apparently not replied after he had sent her a poem. His shot missed and, while police hurried to the scene, two Eton schoolboys – Gordon Chesney Wilson and Leslie Murray Robertson – rained blows down on him with their umbrellas.

Tried for high treason, Maclean was found insane rather than guilty and was dispatched to Broadmoor asylum, where he died some 40 years later. For her part, Victoria was incensed that he could be considered not guilty when she was so obviously and deliberately put in the line of fire. Her ensuing wrath inspired a legal modification that thereafter permitted a verdict of guilty but insane.

The event was immortalised in verse by William McGonagall, acclaimed as the worst of British poets. He wrote:

Maclean aimed at her head;
And he felt very angry
Because he didn't shoot her dead.
Maclean must be a madman,
Which is obvious to be seen,
Or else he wouldn't have tried to shoot
Our most beloved Queen.

Bradshaw's guide bubbles over in its praise of Windsor, claiming its scenery is remarkable for its

> sylvan beauty and the weary citizen who desires to enjoy a summer holiday cannot do better than to procure an admission ticket to Windsor Castle from the printsellers Messrs Colnaghi, of Pall Mall, and then make his way to the Great Western Railway in time for an early train.
>
> Within the next three hours he may see all the regal splendours of the palatial halls of Windsor and then, having refreshed the inward man at any of the hostelries which abound in that town, he may stroll forth into the country and contrast the quiet and enduring charms of nature with the more glittering productions of art with which wealth and power surround themselves.

The interior of the Queen's railway carriage, engraved by Jules David, c. 1844.

It was from neighbouring Slough that Queen Victoria took her first train ride on 13 June 1842 as she journeyed between Windsor Castle and Buckingham Palace. Although she was nervous she was persuaded to make the journey by Prince Albert, who had taken the trip from Slough to Paddington for the first time some three years previously.

With scant notice, Great Western Railways prepared a royal carriage. (It was in fact similar to a standard carriage but was liberally decorated with flowers.) It was placed in the middle of seven carriages – to protect the Queen if there was a shunt – and towed by the locomotive *Phlegethon*. At the controls was locomotive designer Daniel Gooch, who was later knighted for his work laying the first transatlantic telegraph cable by ship, the *Great Eastern*. Alongside him stood Isambard Kingdom Brunel and a royal footman, whose scarlet clothes were soon covered in soot. The journey took 25 minutes. There was a red carpet and a detachment of Royal Irish Hussars at Paddington to meet the Queen, along with an army of GWR dignitaries.

The Queen was now converted to the benefits of rail travel, as a letter written soon afterwards reveals: 'We arrived here [Buckingham

The Queen, a replica of the royal train, at Royal Windsor Station, Windsor.

Palace] yesterday morning, having come by the railroad, from Windsor, in half an hour, free from dust and crowd and heat, and I am quite charmed with it.'

Eventually all the major railway companies produced a royal train, with sumptuous fittings, expressly for use by the royal family, and as the network expanded she was even able to reach her Scottish home, Balmoral, with greater ease. Nonetheless, Victoria applied strict rules to train travel, which were that the locomotive speed did not exceed 40 mph during the day and 30 mph at night, and that the train was stopped at mealtimes.

Of Slough there is not a word in *Bradshaw's* although today the station has a particularly quirky claim to fame in the shape of 'Station Jim', a dog that collected cash for the Great Western Railway Widows and Orphans Fund for two years from 1894. Brought to the station as a puppy, he was taught tricks and barked every time someone popped a coin in the collecting box he wore around his neck. He was credited with collecting more than £40, usually donated in pennies or halfpennies. After his sudden death he was stuffed and placed in a glass case on platform five with a collecting box so he could continue his work posthumously.

Inside the glass case there's an account from the nineteenth century about Station Jim's accomplishments:

> *He would sit up and beg or lie down and 'die', he could make a bow when asked or stand up on his hind legs. He would get up and sit in a chair and look quite at home with a pipe in his mouth and cap on his head. He would express his feelings in a very noisy manner when he heard any music ... If a ladder was placed against the wall he would climb it. He would play leap frog with the boys; he would escort them off the station if told to do so but would never bite them.*

SLOUGH STATION WAS ALSO PIVOTAL IN A SENSATIONAL MURDER CASE THAT CAUGHT THE IMAGINATION OF VICTORIAN ENGLAND

Slough Station was also pivotal in a sensational murder case that caught the imagination of Victorian England, not least for the involvement of a new communication method recently installed there. The murderer was John Tawell, apparently a model citizen and devout Quaker, and his victim was his one-time mistress, Sara Hart. Tawell's chosen method was poison, while the new technology in question was the electric telegraph, which went on to change Victorian communication beyond recognition.

Although Tawell appeared to be a pillar of the community he was in fact a convicted felon. Aged 30, and a husband and father, he was transported to Australia after being found with a forged bank bond. But his initial criminality was laced with charm. Soon he was given his freedom in Australia where he opened a successful drug store selling home-made remedies, proving he was more quack than Quaker. He also surrounded himself with adoring women, all of whom were taken aback when Mrs Tawell and her children arrived.

The couple eventually returned to England, where Mrs Tawell fell ill. Her husband employed Sara Hart to look after their child. Tawell and Hart promptly began an affair. When his wife died they lived together for a while but Tawell soon transferred his affections to another woman, a Quaker in nearby Berkhampstead. Despite his chequered past, the couple were married. However, disquiet among the Quaker community about his previous liaison persuaded him that Sara would have to be eliminated from his life.

The offices of the Electric and International Telegraph Company, in Bell Alley, Moorgate Street, London, 1859.

On New Year's Day 1845 Tawell went to Sara's cottage in Slough and together they shared some ale. Unknown to her, he had poured prussic acid into her glass; the swift-acting toxin left her writhing in agony. A neighbour called the local doctor, who guessed Sara had been poisoned and could do nothing to save her. He did, however, persuade the railway station to send an alert using its newly installed electric telegraph.

At the time the new technology's use was tentative and the number of trained operators few. Fortunately, the far-sighted doctor and the quick-witted stationmaster valued its immediacy. The message sent from Slough read:

> *A murder has gust been committed at Salt Hill and the suspected murderer was seen to take a first class ticket to London by the train which left Slough at 742 he is in the garb of a kwaker with a great coat on which reaches nearly down to his feet he is in the last compartment of the second class compartment.*

(The earliest telegraph kit did not have the letters Q, J or Z, or punctuation, among its keys.) Thanks to the telegraph message, police were waiting for Tawell at Paddington. He was followed and arrested the next day.

The Electric Telegraph

The commercial telegraph was patented by Sir William Fothergill Cooke and Charles Wheatstone in 1837. (Morse code was patented the same year in America although it was some time before the English telegraph and the American dot-dash signal system were united.) The inventors made liberal use of groundwork laid by Michael Faraday, who had linked electrical currents and magnetism and built the first solenoid that combined the potency of both.

Now, using the power of an electric current, magnetic needles could transmit messages in code over long distances. The telegraph was first used on the railway system to link Euston with Camden Town. On 9 April 1839 it was installed along the Great Western Railway between Paddington and West Drayton, and in 1843 it was finally extended to Slough.

Its potential for communication and railway signalling at the time was still largely unappreciated. However, soon the sending of telegrams would be an everyday occurrence for the Victorians, a first sortie into instant communication. Social and business networking was poised to spread around the globe, thanks to the electric telegraph, and by the time the 1866 *Bradshaw's Tourist Handbook* was published, the whereabouts of every telegraph station was included, a reflection of how important the technology had become.

Edison's electric telegraph system for railways, 1886.

Tawell believed his veneer of respectability put him beyond the law. 'You must be mistaken,' he told officers. 'My station in life places me above suspicion.' Soon he claimed Sara committed suicide expressly to implicate him. His wife and some elements of the Quaker community believed he was innocent. However, a London chemist testified at his trial, held at Aylesbury Assizes, that Tawell had bought prussic acid on the day of the murder. And a neighbour said Sara had been ill before following a visit by Tawell.

It was enough to convince the jury that Tawell was a murderer. He was found guilty and hanged at Aylesbury gaol on 28 March 1845 before a 2,000-strong crowd. Before his death he wrote a confession that finally persuaded his loyal wife that he was a killer.

The executioner failed to allow sufficient 'drop' on the rope to cleanly break Tawell's neck and instead he was slowly strangled by the noose. He was, as one newspaper of the day observed, 'the murderer who was hanged by the wires'.

IN SLOUGH THE RAILWAYS BECKONED IN THE FIRST COMMUTER HOUSES IN THE 1840S

In Slough the railways beckoned in the first commuter houses in the 1840s, built at Upton Park by James Bedborough, once a stone mason at Windsor Castle who later established himself as a builder, town councillor and one-time mayor of Windsor. He was clearly a man of vision and ambition, advertising the houses in terms of their proximity to Windsor Castle and the railway.

The new houses bordered parkland that was designed by noted gardener Joseph Paxton, who went on to find fame and a knighthood by designing the Crystal Palace in time for the Great Exhibition in 1851. He also made a fortune by joining the board of the Midland Railway Company.

Joseph Paxton (1803–1865).

Paxton became an MP but never forgot his humble roots. The son of a yeoman farmer, Paxton fared considerably better than Bedborough. Unfortunately, to buy the land in Slough and build the first 29 terraced houses and villas Bedborough steeped himself in debt, and when he died in 1860 there was no money available to settle the accounts. The tangled web apparently led to the suicide of two sons and left his family in financial ruin. He is the ancestor of television presenter Davina McCall.

The railway line between Maidenhead and Slough is distinguished by one of Brunel's most graceful bridges. To build it, he tore up the

Brunel's railway bridge at Maidenhead, built in 1838.

rule books with characteristic flourish and produced the widest, flattest bridge in the world, opened in 1839. Nervous railway bosses who couldn't fathom the design principles insisted he left the wooden framework used during construction in place to help support the bridge. Cannily, Brunel lowered the framework a fraction so it appeared to anxious observers to be bearing a burden. Only when the timber skeleton was washed away in a storm did the apparent miracle of design become obvious.

A decade later the ripples of admiration for this achievement were still fanning outwards. In a county magazine, *Lipscomb's Buckinghamshire*, Brunel's triumph was once again heralded.

> *That this beautiful outline is wholly formed of insignificant little bricks, each course of which on this enormous span has not only to carry its own weight but its proportion of the road and the train. When he considers the strains to which these materials are exposed and remembers that they are subject to a pressure that must approach*

very nearly to the limit of cohesion he will sufficiently appreciate the courage and the capacity which have approached so near to the verge of possibility without transgressing its bounds.

Today the bridge still bears the weight of numerous trains, each ten times heavier than those Brunel would recognize. One of the land arches is known as 'the Sounding Arch' for its remarkable echo.

Isambard Kingdom Brunel (1806–1859) was a major figure in Britain's railway story, particularly on the westbound routes. He was the son of a French engineer, Marc Isambard Brunel, whose own sparkling career was dimmed only by a spell in debtors' prison.

After being educated in Paris, one of young Brunel's first projects was to work alongside his father on a tunnel that went underneath the Thames in London. While he was working there a tidal wave of water broke through the tunnel's defences and left him seriously injured and several workmen dead. As he convalesced he feared for a time that he would only be remembered for a part-built tunnel. He dreaded becoming a 'mediocre success', someone with only measured achievement in his chosen field.

However, opportunity arose even while Brunel's broken leg was still in plaster. There was a competition to find a design for the planned Clifton suspension bridge in Bristol and he submitted four entries. Judge Thomas Telford rejected all competition entries in favour of his own design, but the embarrassed organisers extricated themselves from Telford's plans and held a second competition, which Brunel won. He was still in his twenties, although the bridge wouldn't be built until after his death due to lack of funds.

In 1833 Brunel became chief engineer to the Great Western Railway, which had ambitions to snake into western England through all the major cities that side of the country. To get official permission for the project was no easy task and Brunel was at the forefront of its defence. Although it was thrown out by the House of Lords once, the company's proposal was given the go-ahead on its second presentation in 1835.

On 26 December 1835 Brunel wrote in his journal for the first time in two years:

When I last wrote in this book I was just emerging from obscurity. I had been toiling most unprofitably at numerous things … what a

ABOVE: **Isambard Kingdom Brunel (1806–1859).**

OVERLEAF: **The western entrance of Brunel's Box Tunnel, running under Box Hill between Bath and Chippenham.**

change. The railway is now in progress, the finest work in England – a handsome salary – £2,000 a year – on excellent terms with my directors and all going smoothly, but what a fight we have had – and how near defeat – and what a ruinous defeat it would have been…

The bridge at Maidenhead and the one at Windsor were two in a string of remarkable engineering feats carried out as the line headed west. Another was the Box Tunnel between Bath and Chippenham, some two miles in length, that took almost six years to complete. Two gangs worked inwards from each side of the hill to meet in the middle at almost precisely the point that Brunel predicted – out by only one and a quarter inches.

Brunel was a hard worker who wasn't afraid of getting his hands dirty. However, he had little respect for money and many of the projects he was linked to were overspent. There's ample evidence that Brunel believed his own ideas to be the best, and that he was always the best man for the job. According to the memoirs of Daniel Gooch,

The last broad-gauge train leaving Paddington Station, 20 May 1892.

BRUNEL WAS A HARD WORKER WHO WASN'T AFRAID OF GETTING HIS HANDS DIRTY

published before his death in 1889, 'One feature of Brunel's character ... that gave him a great deal of extra and unnecessary work, was that he fancied that no one could do anything but himself.'

Brunel laid nearly 1,000 miles of track, nearly all of it broad gauge which measured 7 ft. His belief in broad gauge was not necessarily misplaced, as he correctly insisted it would take bigger trains at fast speeds for a more stable and comfortable journey. However, elsewhere in the country narrower gauges were better suited to the terrain, and they were invariably used in railways engineered by Robert Stephenson and his father George. The Stephensons, who did so much for locomotive development and track expansion, favoured a 'standard' gauge of just 4 ft 8½ in. Clearly the same train could not travel on both lines. Passengers whose journey spanned two different systems were compelled to change trains.

Brunel maintained that the Great Western Railway could operate in isolation. But the stark reality was that much more standard gauge track was laid in the early years of railway expansion. Eventually the government insisted on a national standard gauge and the standard it chose was the one favoured by the Stephensons. In 1846 the government made this standard gauge the norm.

For Brunel, the railways formed only part of his vision for extensive travel. Following a train journey he saw the traveller getting aboard a steamer and heading across the Atlantic to New York. He designed the *Great Western* paddle-wheel steamer in 1837, which operated a regular transatlantic service, the first ship of its kind to do so. There followed the *Great Britain* in 1843, the first big ship to use a screw propeller. His last ship was the *Great Eastern*, launched in 1858 after a series of delays. The largest ship ever built at the time, it had capacity for 4,000 passengers and stored sufficient coal to travel non-stop to Australia. But its construction was blighted by bad feeling between Brunel and his co-designer, John Scott Russell, and its first voyage was marred by an explosion below decks. Anxieties surrounding the ship may well have hastened Brunel's death, which came just a month before Robert Stephenson's in 1859.

The success of the *Great Eastern* as a passenger ship was blunted by the speed of smaller ships, which covered the same distances in shorter times. Nonetheless, it earned its place in history in 1863 by laying the first telegraph cable between England and America with Daniel Gooch at the helm.

Brunel's talents were not confined to pushing back the horizons for travellers, though. During the Crimean War in the 1850s he designed camp hospitals to the specifications of British nursing luminary Florence Nightingale, including lavatories and ventilator fans, and these prefabricated modules, put up under the supervision of 18 skilled men, were credited with cutting the number of hospital deaths thanks to this consideration for hygiene. Brunel also designed docks for Cardiff, Brentford and other ports. A noticeably short man, he favoured a stove-pipe hat to improve his physical bearing. His death at the age of 53 was almost certainly caused by overwork.

SEEING ABUNDANT POTENTIAL PALMER'S IDEAS WERE FAR MORE AMBITIOUS THAN ANYTHING THAT HAD GONE BEFORE

Beyond Maidenhead lies Reading, reached by the Great Western Railway in 1840, where the sweet smell of success came from a biscuit factory. And key to the prosperity of the factory was the web of railways that brought the product to tea tables up and down the country.

Huntley & Palmers began life as a small bakery. Joseph Huntley, the son of a headmaster, learned about the business of biscuits from his mother Hannah, who cooked them in the school oven and sold them at the school gates during his childhood. Much later, when he was 51, he moved to Reading and managed sales for his baker son Thomas. As Quakers the family believed in offering quality goods at a fair price. The biscuits were popular and business was good, although there were fewer than a dozen employees on the payroll.

After Joseph retired in 1838, another Quaker, George Palmer, paid £550 to become a partner in the business alongside Thomas. Seeing abundant potential, Palmer's ideas were far more ambitious than anything that had gone before.

He employed a network of salesmen who brought biscuits to new outlets up and down the country. A new factory opened in 1846, dramatically increasing production. Thanks to mechanisation, including automatic biscuit cutters and giant ovens through which biscuit dough was passed, production went up 500 per cent in five years. There was more expansion in the second half of the nineteenth century, incorporating new sites, and the army of employees increased accordingly.

The railways arrived in Reading in 1840 and Palmer had established all elements of the factory close to the railway lines.

Advertisement, c. 1880, featuring a Victorian picnic, complete with Huntley and Palmers biscuits.

Indeed, the factory had its own rails laid to link up the national network, at one point totalling some seven miles. For these lines there were even Huntley & Palmer locomotives, specially made to run on steam so pollution did not taint the biscuits.

The bosses considered themselves strict but fair. There was to be no swearing, no fighting, no drinking or smoking on company premises. Anyone caught contravening the rules was fined. However, all the money generated in this way was put into the Sick Fund, along with a compulsory donation made weekly by each worker. This money helped to finance those struck down by illness, years before state welfare was introduced. The partners regularly held parties for the staff, gave them paid holiday at Christmas, had wedding cakes baked for those getting married and paid for the funerals of those who died in service.

The atmosphere between the partners wasn't always harmonious, though. Thomas Huntley felt the business was growing too fast, too soon, while George Palmer felt he bore the lion's share of commercial responsibility.

In 1857, with 90 men now baking biscuits, Thomas Huntley died, leaving a son, Henry, who had no burning interest in the business. The Huntley and Palmer families parted company, with a sum of almost £34,000 being paid to the outgoing Henry Huntley. George Palmer's two brothers, William Isaac and Samuel, became partners instead.

By 1894 there were 5,000 people employed by the Huntley & Palmers factory working a 54-hour week, including unmarried women who were segregated from the men. (It would be another 50 years before the company agreed to employ married women.) Biscuit making was so key to the town that the prison was known locally as 'the biscuit factory' and the local football players were called 'the biscuit men'. Rows of red-brick houses sprang up around Reading to provide accommodation for the biscuit factory workers. Meanwhile, packets of biscuits in 400 varieties were being dispatched to 137 countries by 1904, with the first leg of their journey undertaken by rail. The shape of wholesale biscuit tins was even altered to better fit into railway carriages.

From Reading via a Great Western branch line the Victorian traveller heading south could reach Basingstoke – a 'straggling, ill-built town', according to *Bradshaw's* – to be united with the South Western main line. Basingstoke was for a while the end of the South Western line. It was finally linked with Winchester after four tunnels and a station were constructed and the line between London and Southampton was completed in 1840. Its first train took three hours to cover a distance of just under 80 miles between its terminus at Nine Elms and the end of the line at the south coast. Had the government been involved, the railway's destination would no doubt have been Portsmouth, where the Royal Navy had a large and important base, but the London & South Western Railway Company chose Southampton above its neighbour, although its dock facilities were modest.

BY 1894 THERE WERE 5,000 PEOPLE EMPLOYED BY THE HUNTLEY & PALMERS FACTORY WORKING A 54-HOUR WEEK

From Basingstoke the line forges through some sparsely populated countryside before glancing off the western edge of the South Downs. On the face of it, Micheldever Station on the South Western line was so insignificant it did not warrant a mention in Bradshaw's guide. Indeed, it is unlikely that *Bradshaw's* could have ever envisaged the reason for its fame. However, in 1895 Micheldever Station was the unlikely starting point of the first ever car journey in the UK.

The car was a Panhard-Levassor powered by a Daimler engine and it arrived by rail from France, where cars were a far more common sight, via Southampton docks. It was then transported to the Hampshire Station by rail. At the time it cost £200, which equates to about £14,500 in today's money. The purchaser was the Honourable Evelyn Ellis. He was accompanied on its first journey by engineer Frederick Simms, who wrote an article for the *Saturday Review* on the adventure.

> *We set forth at exactly 9.26 a.m. and made good progress on the well-made old London coaching road; it was delightful travelling on that fine summer morning. We were not without anxiety as to how the horses we might meet would behave towards their new rivals, but they took it very well and out of 133 horses we passed only two little ponies did not seem to appreciate the innovation. On our way we passed a great many vehicles of all kinds [i.e. horse-drawn], as well as cyclists. It was a very pleasing sensation to go along the delightful*

LEFT: **The Hon. Evelyn Ellis in his Panhard-Levassor, 1895.**

OVERLEAF: ***Bodmin*, a steam locomotive pulling on the Watercress Line.**

roads towards Virginia Water at speeds varying from three to twenty miles per hour, and our iron horse behaved splendidly. There we took our luncheon and fed our engine with a little oil. Going down the steep hill leading to Windsor we passed through Datchet and arrived right in front of the entrance hall of Mr Ellis's house at Datchet at 5.40, thus completing our most enjoyable journey of 56 miles, the first ever made by a petroleum motor carriage in this country, in 5 hours 32 minutes, exclusive of stoppages and at an average speed of 9.84 mph.

The trip was more daring than it first appears. There were of course no petrol stations, so the oil they bought was from a chemist's shop. But Ellis and Simms were also flouting the law, which decreed motorised vehicles could only use the roads if a man walked in front waving a red flag. This law was repealed the following year. Still, on its early performance there was little to suggest that the motor car would have overwhelmed the train in terms of popularity and usage within a century.

It would be decades before road transport made it impact on industry in Britain. In the meantime, businesses that relied on getting produce to consumers in timely fashion were flourishing, as the railway network continued to expand.

The watercress business fell into this category, and its success contributed to an improvement in the general health of Victorian society. Watercress was favoured by rich and poor alike. Children would pluck bunches of the naturally occurring vitamin-rich plant to eat in the hand. Aristocrats enjoyed it as it garnished silver-service meals on porcelain plates. Its health benefits were immense, as it contains iron, calcium, sulphur and other nutrients.

Watercress was first farmed in Britain at the beginning of the nineteenth century. Hampshire soon became a centre of the trade as crops flourished in the chalky waters of its rivers. Before trains reached the county, stagecoaches were used to transport it to London. When the county was finally fractured by rail services, the one serving the heart of the industry was nicknamed 'the watercress line'. It was built between the existing stations at Alton and Winchester and was officially called the Mid-Hants Railway after it opened in 1865. The stations on the line were Itchen Abbas, Ropley and Alresford. Another station, Medstead and Four Marks, was opened three years later.

After willow flats or hampers filled with watercress were loaded onto the trains, they were taken to London and then across Britain for the watercress to be sold as a low-cost snack. Always a secondary line as far as the railway authorities were concerned, the Mid-Hants Railway nevertheless attracted interest for combating some severe gradients, ascending at its highest point to 652 ft (199 m) above sea level.

When the railway line linking Winchester and Southampton was built in 1838 by the London & South Western Railway, Winchester was its temporary terminus. Eventually tens of thousands of visitors used the train to wonder at the ancient glories in the Roman city. *Bradshaw's* describes how one of the first organs made in England was installed in Winchester's cathedral in the year 951. 'It was a ponderous thing containing 400 pipes blown by 24 pairs of bellows.' In 1854 the organ was replaced by one that had thrilled visitors to the Great Exhibition three years earlier, with 5,500 pipes.

At first, Winchester was the only major centre of population on the London to Southampton line, with Basingstoke a minor market town. The railway's arrival altered the destiny of one hamlet along its route, transforming it into a major industrial hub.

Although Eastley had existed for a long time, it was so remote that the station built there was called Bishopstoke Junction in order that passengers might better recognise their whereabouts. Early on, the railway company built cottages there to house staff, and it assumed greater importance when it was linked by a branch line to Portsmouth. From 1852 it was the site of a cheese market and even had a dedicated siding for cheese transport.

A programme of house building finally united two parishes, Barton and Eastley. From 1868 the combined parish was called Eastleigh at the suggestion of a local author, Charlotte Yonge, who donated £500 for the building of a parish church. Growth accelerated still more after 1891 when the London & South Western Railway opted to switch its carriage and wagon works from Nine Elms – previously the company's London terminus – to Eastleigh and, later, its locomotive works. The site wasn't picked for its proximity to coal deposits or iron ore mining works but for its unexpectedly strategic position on the company network.

By the end of the Victorian era Eastleigh was known as a railway town, with some 2,600 people in the pay of the railway company. It

was also a passing point for military traffic during the Boer War, all funnelled overseas through Southampton. Much later, commentator John Arlott said: 'The very name "Eastleigh" means railways … Eastleigh has a heart – a huge fiery, steam-pulsed, hammer-beating heart.'

While the British love affair with railways continued, the economy of Eastleigh was assured. But, like other towns, it was at risk when passenger numbers declined. Eastleigh has been comparatively lucky, with some locomotive work still continuing today in the town, although on a much-reduced scale.

'THE VERY NAME "EASTLEIGH" MEANS RAILWAYS … EASTLEIGH HAS A HEART – A HUGE FIERY, STEAM-PULSED, HAMMER-BEATING HEART'

Queen Victoria became well acquainted with the railways of southern Britain, and especially those of the L&SWR, as she made regular trips to Osborne House on the Isle of Wight, apparently her favourite royal residence. Her patronage of the island in turn led to a major boost in its tourism.

Today a visitor on the boat trip from Southampton would feel hemmed in by the commercial dockyards and colossal container ships that line the water's edge. For Queen Victoria the view would have been very different. According to *Bradshaw's*, Southampton was

> 79 miles from London by the South West Railway, on a point at the head of a fine inlet called Southampton Water into which the [rivers] Test and Itchin run.
>
> When the mudbanks are covered at high tide its inlet is a fine sheet of water seven miles long and one or two broad, and exactly the spot for a sail with groves along the shores, especially in the west, in which the nightingales are heard all night long.

When Queen Victoria disembarked from the ferry she might well have fancied a spot of lunch, and the prestigious yacht club at Cowes was the perfect destination. There was one problem: the club only allowed men inside its doors. The single-sex ruling would not be relaxed even for the Queen of England. However, the club was obviously keen to garner royal custom so a conservatory was built leading from the club

PREVIOUS PAGE: Visiting Winchester Cathedral.

LEFT: The yacht *America* won the sailing match at Cowes for the Club Cup – the race was renamed 'The America's Cup' in her honour.

but, as far as etiquette was concerned, not a part of it, and here the Queen was welcome.

Queen Victoria and Prince Albert bought Osborne House in 1845 for £28,000. They didn't want a palace – they had several of those already – but a family home that would be a refuge from the demands of public life for themselves and their children.

The house they bought was grand but unexceptional. Prince Albert set about redesigning the building and favoured Italian architecture that he had seen on his travels prior to marriage. He collaborated with Thomas Cubitt, an eminent builder whose company had been responsible for many of London's most prestigious and stylish neighbourhoods. The effect of their plans was startling and sympathetic. The attention to detail extended to the interior, to the luxurious fixtures and fittings bearing Victoria and Albert's monograms, carefully chosen by the royal couple to complement their island home.

Osborne House gave the royal family an opportunity to enjoy a brand of domesticity, albeit one that was very much sheltered from the stresses of everyday life. In 17 years Victoria gave birth to nine children: Victoria, Edward, Alice, Alfred, Helena, Louise, Arthur, Leopold and Beatrice. It was in many ways a remarkable feat as Victoria was ambivalent about

ABOVE: Osborne House, Queen Victoria's home on the Isle of Wight.

OVERLEAF: Alum Bay, Isle of Wight.

the young. 'I don't dislike babies,' she once insisted, 'though I think very young ones rather disgusting.' Included in the grounds of Osborne House was a chalet known as the Swiss Cottage, where the couple's brood were educated. Beyond the tutoring that many children of the era came to expect, Prince Albert insisted they learned how to grow vegetables and, further, to sell and cook them.

When Albert died after contracting typhoid in 1861 at the age of just 42, Victoria turned her room at Osborne into a shrine dedicated to him. There was a portrait of her husband set up by her bed so it was the first thing she saw every morning. His pocket watch, regularly wound, was kept in a pouch at the bed head. And when Victoria herself died at Osborne House in 1901 after suffering a stroke, the pattern was repeated, her son Edward ordering metal gates to be fitted in the corridors so her room would not be disturbed for 50 years. It wasn't until 1954 that Queen Victoria's great-great-granddaughter Queen Elizabeth II consented to the gates being opened.

Queen Victoria's body was taken from the south coast to London for the funeral service. Afterwards her coffin was taken back to Windsor by train for burial.

AFTER IT BECAME THE FOCUS OF ROYAL FAVOUR, THE ISLE OF WIGHT ENJOYED SOMETHING OF A HEYDAY AS TOURISTS FLOCKED THERE

After it became the focus of royal favour, the Isle of Wight enjoyed something of a heyday as tourists flocked there. Among the attractions were the variously coloured sands of Alum Bay. The colours, which form naturally in the cliffs, were used either to create decorative strata in glass jars or, more intriguingly, to make pictures.

Sand art was pioneered by German painter Benjamin Zobel in the eighteenth century when he was a court artist at Windsor Castle. Before his tenure, the royals were amused by pictures made from sand which were, by their nature, temporary. However, using gum Arabic and white lead, Zobel found himself able to stick the sand to a board to create more lasting images, calling the results of his technique 'marmotinto'. During Victorian times this style of art enjoyed remarkable popularity, particularly on the Isle of Wight where visitors collected the coloured sand from the cliff face.

With rail travel the Victorians could indulge their curiosity as never before. They had a weakness for freakiness and it was this fancy for the strange, shocking and downright extraordinary that brought fame to Joseph Merrick as the Elephant Man (and to any number of excessively tall, short, fat, thin, physically tormented or unusually hirsute people). Previously their curiosity was confined to sensational pamphlets or dependent on visiting circuses or neighbourhood 'penny gaffs', informal venues that hosted travelling shows. However, thanks to an ever-extending network of trains, the upper- and middle-class Victorians now travelled to marvel before novelties and oddities at work and play. For those who became spectacles the dubious reward of celebrity was rarely matched with fortune.

Beyond the New Forest on the south coast lies Bournemouth, described by *Bradshaw's* as

> a fashionable modern watering-place and winter residence ... It is situated in a beautiful sheltered spot in the chine of low chalk cliffs and is much resorted to by invalids for its health situation and quiet retirement.

Some of Bournemouth's health-giving qualities were attributed to the great number of pine trees planted by the town's first landowners in the early nineteenth century. There is evidence of this in street names that still exist today, including Pine Walk, which was originally called Invalids' Walk by the Victorians.

Settlement didn't start in earnest until Queen Victoria acceded to the throne, so Bournemouth became a truly Victorian town. The town's first hotel, the Royal Bath, opened on the day of the Queen's coronation: 28 June 1838. Among the notable visitors to have stayed there are Prime Ministers John Russell, Benjamin Disraeli and William Gladstone, Empress Eugenie, the wife of Napoleon III, and other foreign royalty.

Given the town's glorious reputation for good health, the train, when it finally arrived in Bournemouth, led to an influx that more than doubled its population in a decade. Still, Bournemouth was slow out of the blocks. An extension to a branch line serving Ringwood and Christchurch arrived in 1870. In 1874 Bournemouth West Station was opened for the Somerset & Dorset Joint Railway, although passengers were compelled to change at Wimborne. Two years after that it was linked to Poole by rail. Only in the 1880s did the improved railway system begin to have an impact on visitor numbers. By the time Victoria died Bournemouth encompassed some of its surrounding villages and had a population of 59,000. It also had by then a library, a symphony orchestra, a hospital and a pier.

For the Victorian visitor guided by *Bradshaw's*, making a choice between Bournemouth and its near neighbour Weymouth was not difficult. Bournemouth is dismissed in little more than a line. Weymouth, on the other hand, is given almost ceaseless praise for being 'striking', 'picturesque' and 'delightful'.

> No place can be more salubrious than Weymouth. The air is so pure and mild, that the town is not only frequented during the summer but has been selected by many opulent families as a permanent residence; and the advantages which it possesses in the excellence of its bay, the beauty of its scenery and the healthfulness of its climate, have contributed to

raise it from the low state into which it had fallen from the depression of its commerce, to one of the most flourishing towns in the kingdom.

Points of interest included a regatta, a theatre, assembly rooms, a castle and pleasure boats. Yet most went to bathe in the crisply chill waters off the Dorset coast. To enter the water, swimmers climbed into a bathing machine, divested themselves of clothes and were towed into the water either by horse- or manpower. At a certain depth they could modestly plunge into the water away from prying eyes. The machine was left in place until the dip was finished so bathers could return to shore the same way, after waving a flag to attract the attention of the machine driver.

When bathing became fashionable in Victorian times the first swimmers were men and they went into the water naked. Costumes became compulsory for all by the 1860s, not least to spare the blushes of women swimmers, and by the 1880s swimsuits had become suitably large and robust so that women were unafraid to be seen in them. The era of bathing machines was at an end. But when *Bradshaw's* guide was written there were still two decades to go before swimmers strolled unaccompanied into the sea from the shore.

The water is generally very calm and transparent. The sands are smooth, firm and level and so gradual is the descent towards the sea that, at the distance of 100 yards, the water is not more than two feet deep. Bathing machines of the usual number and variety are in constant attendance and on the South Parade is an establishment of hot salt-water baths furnished with dressing rooms and every requisite accommodation.

Weymouth was also at the end of the Somerset, Wiltshire & Weymouth Railway, planned from 1845 but not completed until 1857. This was a delightful holiday line edging down from Bristol and Bath and prized by the Great Western Railway, which took over the line, for its cross-Channel port.

Bradshaw's discusses the merits of Portland, linked to the coast by a ridge of shingle called Chesil Beach, in terms of a day trip from

PREVIOUS PAGE: A view of Bournemouth from the West Cliff.

LEFT: Queen Victoria's bathing machine on the Osborne House estate.

Weymouth. It fails to note the impact of industry on the peninsula which sets it apart from resorts around it. Portland was among the forerunners of the rail age with the Merchant Railway, built in 1826 with a unique 4 ft 6 in gauge. It's not lodged into the history of railways because, with innovation coming thick and fast, its owners failed to keep pace with the changes.

The Dorset line launched in the wake of the pioneering Stockton to Darlington railway on which George Stephenson's *Locomotion* had towed cargo and passengers. Already industrial tracks and waggonways were relatively common sights. And thanks to the Industrial Revolution, the makers of the Merchant Railway could lay iron rails rather than wooden ones. This reduced friction and allowed for longer, smoother journeys.

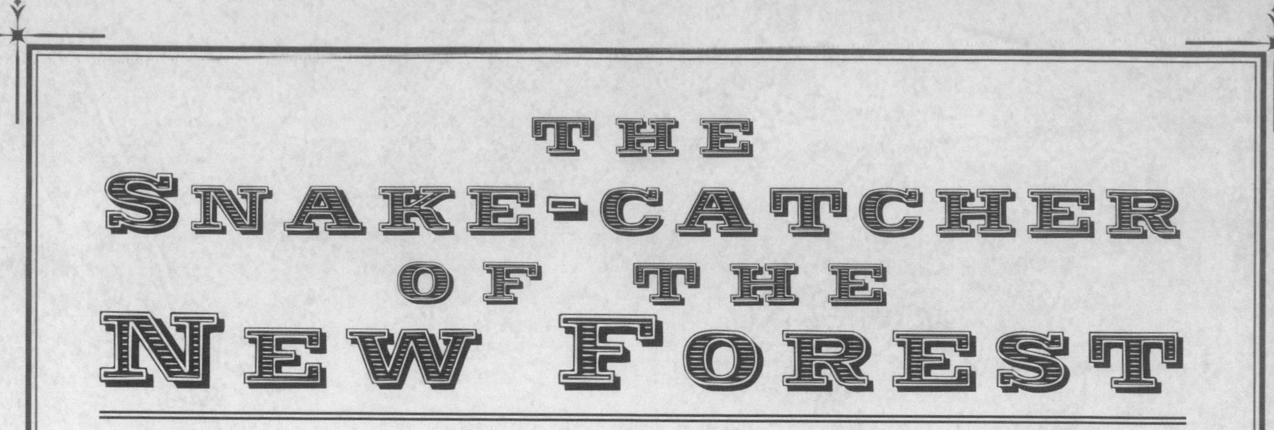

One man who capitalised on the vogue for quirky tourism was Harry 'Brusher' Mills, the famous snake-catcher of the New Forest. This charismatic countryman was happy to share the secrets of his success with hordes of visitors who descended on the Brockenhurst station after it opened for just a few pennies.

Places like the New Forest were populated with snakes to a far greater extent than now, including poisonous adders. Using his bare hands, Mills caught snakes, killed them and made up snake-bite antidote from their body parts. He even laid claim to a cure for rheumatism from the cooked-up bodies of New Forest adders. Dead snakes were also transported back to London by train to feed animals at London Zoo.

With a forked grey beard, Mills cut an eccentric figure. Whatever the weather he wore two coats over a waistcoat and gaiters. His snake-catching kit included sacks, a pronged stick and a sharp knife. He used this to make a deep cut in his own flesh if he ever suffered a snake bite, before applying a home-made ointment. It's thought his snake tally eventually numbered something around 30,000, including some 4,000 adders.

Although he was hampered by a cleft palate it never stopped him from chatting with visitors. They were welcomed to his hand-built hut for a cup of tea, brewed on a wood fire, served in tins without milk but laced with sugar and stirred by a homemade spoon.

An ancient law governing the New Forest would have given him 'squatters' rights' where he lived if his hut had not been burned down one night, apparently by vandals. Stories abound about Mills, who was known to clear a path to the bar through a crowded pub by casually tossing a live snake on to the floor.

HIS SNAKE-CATCHING KIT INCLUDED SACKS, A PRONGED STICK AND A SHARP KNIFE

Still he was clearly held in fond regard by locals, not least for establishing the New Forest as a must-see tourist attraction. After his death a collection among friends and drinking partners raised sufficient money for a marble memorial in the churchyard at Brockenhurst. It bears the following inscription:

***His pursuit and the primitive way
in which he lived, caused him to
be an object of interest to many
He died suddenly July 1st 1905
aged 65 years.***

WITH CHARACTERISTIC VISION, THE VICTORIANS SAW THAT PORTLAND HAD POTENTIAL AS THE LARGEST DEEP-WATER HARBOUR IN EUROPE

Perhaps because it was built on a one in seven incline – from the quarry at the hill top to the port below – the Merchant Railway failed to graduate to locomotive traffic like other early railways. Instead it depended on either horsepower or chains and wire hawsers to haul its wagons. As a consequence it stayed isolated on the end of the Portland peninsula, unable to expand in the same way as other railway lines. Nonetheless, the Merchant Railway remained in service until the start of the Second World War.

Its freight was soft, grey Portland stone, quarried for centuries and already apparent in St Paul's Cathedral and Buckingham Palace. With the Merchant Railway, however, some very tricky transport issues were resolved for the industry. With stone from the quarries now dispatched quickly to the port side and winched aboard cargo ships, thanks to the railway the market for Portland stone was dramatically expanded.

Meanwhile, an ambitious Victorian plan for Portland would keep quarrymen busy for decades. Portland was already a frequently used harbour for shipping but it was known for its treacherous waters. With a near-perpetual meeting of tides, the offshore Portland Race was a hazard that hampered the progress of the port.

With characteristic vision, the Victorians saw that Portland had potential as the largest deep-water harbour in Europe, key for a country that invested as heavily in its navy as Great Britain. Several breakwaters were needed to create a refuge from the elements. Portland not only had a suitable seafront but an abundance of breakwater stone. With plenty of convicts held in the nearby Portland prison, thanks to a tough national legal code, there was also a labour force that would oblige in the mammoth building project.

Prince Albert laid the first stone in 1848, signalling the start of a quarter-century-long project. An estimated 5,731,376 tons of stone were taken from Portland quarries out to sea. (Subsequently there was still sufficient to provide a tombstone for every Commonwealth grave in the two world wars and for the Cenotaph in Whitehall.) It was tough and treacherous work for the convicts, with one estimate claiming that a dozen were killed for every year it was under construction. However, locomotives operating on broad-gauge track built for the purpose, in addition to the Merchant Railway, made the labouring inestimably easier.

In 1858, a full decade after the breakwater project began, writer Charles Dickens wrote about what he saw when he visited the site in his own magazine, *Household Words*.

Up the hill to the right run the inclines; the heavy four-wagon trains rattle down them and flit by us, each with 'Prince Albert' or 'Prince Alfred' puffing away behind and dashing them off rapidly to the far end of the cage...

A good railed passage is provided, leading between two of the five broad-gauge roads which run to the end of the inner breakwater abreast over open rafters. The large blocks of heaped stone, which at first underlie the rafters, soon become dashed with surf and then give way entirely to the sea, which, if the day be at all fresh, will give the visitor a sprinkling. Six hundred yards from the shore the inner breakwater ends in a noble bastion-like head rising with smooth, round sides some 30 ft above the waves. A space of four hundred feet separates this head from its partner, the precisely similar work at the end of the outer breakwater...

It is a scene of bustle. Here, we pass a gang of men preparing timber for the shores and brackets that support the road-pieces; there, we see a man running along the narrow footway of the workmen – a single plank laid on each side of the rails – as much at ease as if a false step would not tumble him thirty feet down into the sea or worse upon the rugged rubbly heap; which, now emerging from the waves, indicates what the nature of this outer arm is hereafter to be...

Every two or three minutes comes rumbling behind us a train with its four loaded wagons, each wagon averaging 12 tons in weight. An ordinary load consists of a large block in the centre, some two or three feet in diameter, around which are heaped fragments of smaller sizes, the whole rising to a considerable height in the wagon. It is a fine thing to watch the tipping of the rubble through the open rafters of the cage. Every wagon has a dropping floor, slanting downwards from back to front, but with its iron-work lighter and less massive in front than behind. It is so contrived that a brakesman, with a few blows of his hammer, knocks away the check and sets the floor free to drop ... A puff or two of the engine brings each wagon in succession over the required spot and, unless the large stone should become jammed, the whole load is tipped and the empty train is on its way back in less than a minute.

ABOVE: **The Verne, Portland, Dorset.**

OVERLEAF: **Her Majesty's Prison, Portland, on Portland Bill in Dorset.**

The jamming, when it happens, is an awkward business, and men are sometimes at work for hours with picks and crowbars before some obstinate mass will slip between the iron sides. Such accidents are almost always the result of careless packing on the part of the convicts at the top of the inclines, the process being, indeed, one that demands not a little art and skill.

Convicts were also employed in another major undertaking, as the limestone landscape encasing Portland was forever altered by the building of Verne Citadel. This impressive fortress, made once again from Portland stone, initially housed prisoners building the breakwaters. It was later given over to the army and could accommodate 700 men in peacetime, and at least double that in times of war. It was ideally placed to guard the new Portland harbour that opened in 1872. (Two further breakwaters were added in the early twentieth century.)

The effect of all this on native Portlanders can only be guessed. Novelist Thomas Hardy alluded to Portland in his 1897 work *The Well-Beloved* as the Isle of Slingers for the residents' accuracy at throwing stones, a skill honed to keep strangers at bay. Incomers were referred to by local people as 'kimberlins', and no one was permitted to use the word 'rabbit' for fear of invoking a quarry collapse in the same way as rabbit burrows were wont to do.

However, a liking for eighteenth-century life did nothing to stop Portland being dragged into the twentieth century. The creation of the harbour led to trials of self-propelled torpedoes, a new weapon created by British engineer Robert Whitehead in 1866. In Britain the concept of stealthy torpedoes was considered unsporting in the late nineteenth century, so Whitehead manufactured torpedoes overseas for a worldwide market. But ultimately Britain, like every other country, needed an arsenal incorporating every conceivable weapon as the largely peaceful Victorian age gave way to periods of unprecedented global conflict.

As international tensions increased the Admiralty Board quietly told him they would only purchase weapons made in Britain. The venue for Whitehead's new factory was obvious: Portland was already a tried and trusted environment as far as he was concerned. It often hosted Royal Navy ships and there was plenty of space available for a factory on the harbour shore. Thus Britain's first torpedo manufacturer opened at Ferry Bridge in the parish of Wyke Regis, providing new job opportunities for local people.

Even in 1914 torpedoes were viewed with suspicion, with the military 'old school' regarding them in the same way Queen Victoria first saw trains. That year's *Naval Review* revealed the continuing caution when it stated:

> *Until lately it was a generally accepted maxim that the torpedo could not play any part in a fleet action until one side had established a definite superiority in gun-fire and that then its function was merely to complete, in the shortest possible time, the work begun by the guns.*
>
> *The introduction of the long range torpedo has changed these conditions entirely and rather suddenly. Its range approaches equality and its effective range may in some circumstances prove superior to that of the gun; it is quite conceivable that some future fleet actions may commence with torpedo fire.*

JOURNEY
4
Border
Country
Berwick-upon-Tweed
to the
ISLE OF MAN

Railways arrived hand in glove with immense and spectacular bridges that became enduring monuments to the triumph that was British engineering. One of the most striking is the Royal Border Bridge, built by Robert Stephenson in 1850, one of 110 bridges needed for the newly constructed York, Newcastle & Berwick line. It has 28 arches – 15 over land and 13 over water – made from brick, clad with stone. A steam-driven pile-driver pioneered by engineer James Nasmyth in 1845 forced girders into the dense gravel on the floor of the River Tweed to a depth of some 36 feet to ensure the foundations were secure.

Stephenson made such a good job of building the bridge that it didn't need serious repairs until the 1990s. And although it didn't precisely mark the England–Scotland border, as its name implied, Queen Victoria cannot but have been impressed when she officially opened the bridge in August 1850.

But Stephenson had good reason for making this bridge soundly. He was the engineer behind the ill-fated bridge over the River Dee that had collapsed under the weight of a train in 1847, with the loss of five lives. The reason for this catastrophe is still not entirely certain. The disaster happened within a few hours of ballast being laid around the rails by way of fire prevention, following a blaze in Uxbridge, London, which had caused a bridge built by Isambard Kingdom Brunel to crumple. Some believed the ballast brought extra weight to bear on the bridge. Others thought the ballast may have caused the train to derail. Doubts were also raised over the dovetailed cast-iron girders that had been used on the Dee crossing to hold up the Chester to Holyhead railway line. Beforehand there had been warnings issued about the possible hazards of using cast and wrought iron in bridges.

Stephenson was accused of negligence during a later investigation by the Railway Inspectorate, formed in 1840 to safeguard passengers. He was also quizzed closely by a subsequent Royal Commission. He felt the engineering failure keenly. However, most engineers of the day had no wish to see an accomplished man like Stephenson publicly made a scapegoat. Many agreed that virtually all engineers of the era were guilty of being short-sighted about the viability of cast-iron girders.

It is a tribute to his strength of character, and a sign of the bold times, that Stephenson's career did not end in tatters with that

PREVIOUS PAGE: The High Level Bridge at Newcastle, designed by Robert Stephenson.

RIGHT: *The Illustrated London News* reports on the Dee Bridge disaster.

OVERLEAF: Stephenson's Royal Border Bridge at Berwick-upon-Tweed.

THE LATE RAILWAY ACCIDENT, AT CHESTER.

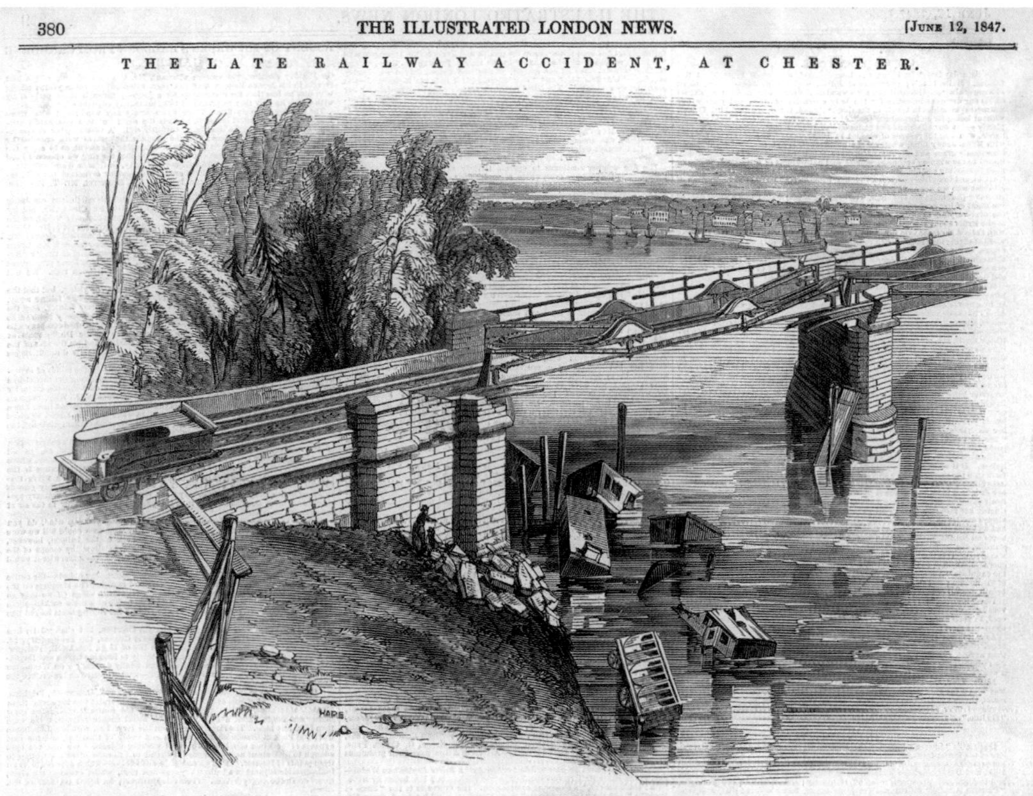

SCENE OF THE LATE RAILWAY ACCIDENT, AT CHESTER.—DILAPIDATED SPAN OF THE DEE BRIDGE.

IN fulfilment of our engagement, we this week present our readers with Illustrations descriptive of the late lamentable accident at the Dee Bridge, on the Chester and Holyhead Railway.

The general view is taken from the high ground on the Saltney side, looking down upon the dilapidated span of the Bridge, and showing the space left open between the piers by the fall of the girder and road-way. The bent ends of the overhanging rails are shown for the purpose of marking more clearly the late track; the rails, as well as every other part of the ruin having been removed from the verge of the opening before we visited the spot.

The Bridge crosses the river at an angle of about 48°, is constructed with three spans—skewed to the same angle—of 100 feet each; each span being sustained by four trussed girders, one on each side, and two in the middle, making the two roadways independent of each other; on the inside of the bottom flange of each pair of girders, shoes are cast, having a doon-tail socket, into which wrought-iron ties are fitted to secure the girders from springing outwards at the bottom, a tendency to which is occasioned by the weight of the road-way and the oscillating pressure of the passing trains. Between these, and resting upon the same flange, are strong timber bearers or joists, upon which a flooring of four-inch planks is laid; on this the longitudinal sleepers are fitted, carrying the rails and check-rails, the latter being continued twenty-six feet beyond the span of the Bridge each way. Between each pair of girders near twenty tons of ballast had been recently laid, and we were informed by a gentleman on whom we can rely, that the unfortunate train in question was the first that had attempted to cross the Bridge after the ballast was so deposited.

Having heard much about the deflection of the girders when a train passed over, we watched them carefully on the occasion of two goods trains coming up, and could not perceive any more than a slight vibration, certainly nothing like a deflection of *inches*; they were unaccompanied by engine and tender. We have given this brief and general description, with the view of making the following details—more immediately referring to the melancholy catastrophe—clear and intelligible to the general reader.

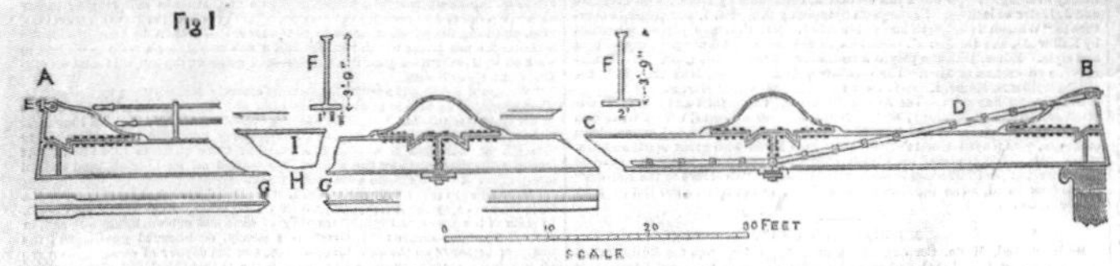

ELEVATION OF GIRDER.

By the evidence given on the inquest, the public are already aware that the same facts are adduced in support of widely differing opinions; and—as it is our object rather to furnish the material for others to judge from, than to volunteer an opinion of our own—we shall proceed to state the facts as they existed on our visiting the spot.

Fig. 1 is a side elevation of the broken girder, showing the exact form and position of the fractures. A is the Saltney end. From B to the fracture C, the girder is represented in its perfect state (except the rails which are indicated at the opposite end, and shown entire in the general view), with the truss or tension plates, D, which run through, and are secured at A and B to the plates E. F F are enlarged transverse sections, and G G are plans of flange, showing the fracture, H. At the end B, is a section of masonry, showing the bearing of girder.

Fig. 2, is an elevation of the inside of the parapet, commencing from the end of the fallen girder, and extending 26 feet toward the Saltney end; the fractures and abrasions are carefully marked; as, on this piece of shattered wall, arguments of a very varied character are founded. J shows the end of the girder; K, where the tender first struck the wall; L L L, marks of abrasion, made by the screws and other slight projections on the side of the tender, several of which are broken off, and others much ground down.

The train consisted of the engine and tender, following which the carriages were arranged—1st. One first-class. 2nd. One second class (with break and guard-box): 3rd. One second: 4th. Luggage-van: 5th. Second-class.

One opinion, having the weight of high authority, is, that the tender, by some means, got off the rails whilst upon the Bridge, and struck the girder at I., which, instantly giving way, the train fell through, the connection between the first-class carriage and the tender being broken in the fall: whilst the latter, having received additional impulse from the engineer turning on the steam just at this point, cleared the Bridge, struck and grazed the wall as before-named, running, still off the rails, a distance of seventy feet, to where it now lies. The engine had parted from the tender, and continued on the rails, having sustained scarcely the slightest injury.

Another opinion attributes the accident to some radical defect in the material or construction of the girder. Evidence having been given that the deflection had amounted to from four to six inches during the passage of a heavy train, it is inferred that on this occasion the girder gave way first at C., and that the piece I. was "jumped" out by the fall.

It was at this point that the engineer states he felt the sinking, and turned the steam full on; the sudden jerk from which gave an impetus to the tender, and enabled it to reach the Saltney side. The way in which it got off the line, before it reached the end of the check rails, is thus accounted for:—

The tender has six wheels: the curvature of the sinking rails would throw two or four of them out of bearing, where the slightest impediment or impulse, at either side, would make it change position on the line; and that this occurred just before striking K., and that the coupling at *a* was broken by the sudden elevation of the fore part, and consequent bending down of the screw to the tug of the engine.

The evidence of the boy went to show that the carriages had nearly all crossed the Bridge, before the entire floor and rails gave way, and that the last carriage ran back, dragging the others, which had become disengaged from the tender, into the river; and it does appear that the first-class carriage struck the parapet at K, from whence portions of the plate glass and window frame were projected, and were found, and seen by us, on the 5th June, lying on the coping stone, as marked in the large drawing at X. From this, it would appear that a portion of the carriage was *outside* the parapet at the time it struck, which would make it very unlikely that any of the carriages had actually reached the Saltney side in comparative safety.

Another opinion is that the masonry was defective, and that the girders had not sufficient bearing; but, on examination, no deficiency was apparent in either respect. Others, that the tender struck the end of the girder A, and dislodged it from its bearings, from whence it fell into the river, and got broke by the fall ;but there is no appearance on the coping of its having been driven off in that way the stone being perfectly free from any marks of the girder having slid *outward*.

There appears no very clear mode of accounting for the tender leaving the line where the check-rails were fixed, except by supposing a curve in the rails; and, if a curve did occur, it must have been produced by something having sunk or given way.

The conduct of the engineer on the trying occasion is deserving of all praise; and there is a satisfaction in knowing that all that presence of mind and courage could do, was done at the *moment*, and done well.

Whilst we regret the absence from Chester of Mr. Lee, the Engineer to the line, and Mr. Munt, of the Chester and Holyhead Railway, to whom we were directed for information, we gladly acknowledge the readiness with which those gentlemen who were on the spot, and connected with the Railway, furnished us with some of the foregoing facts; and, if we have omitted any points which to them seemed important, it is only because we wished to avoid implying censure on any parties.

PARAPET OF THE BRIDGE.

ABOVE: George Stephenson (1781–1848), the great English railway engineer and father of Robert.

LEFT: The High Level and swing bridges at Newcastle upon Tyne, 1890.

OVERLEAF: At the Royal Border Bridge.

calamity. In fact it continued with a pace that matched one of his own locomotives. The Newcastle to Berwick line bears Stephenson's stamp elsewhere too.

In Newcastle the High Level Bridge was opened a year before the one at Berwick, and is equally treasured today. Running 100 feet above the waters of the Tyne, it was neatly dual purpose with a road running underneath the rail bridge. Bridging the river it also linked two communities that had been hitherto kept separate. Newcastle's elegantly curved railway station was also influenced by Stephenson. However, it is perhaps worth noting that his projects helped to obliterate ancient castles in both Berwick and Newcastle.

Stephenson was already an immensely important figure in the north east before the advent of the Royal Border Bridge. His name was over the door of the family engineering works in Newcastle where locomotives were made, an early sign of the faith that his father George invested in him.

George Stephenson died two years before the opening of the Royal Border Bridge, but he had been a huge influence on Robert and no doubt a steadying one after the Dee Bridge collapse. To Robert he bequeathed a singular belief that the future lay with locomotives no matter what technical hitches lay in the way. George was a cow-herder turned coal-picker who didn't learn to read and write until he was in his late teens. But he became one of the most famous engineers ever known, credited with levering Britain into the modern age.

His fascination for the early steam-powered engines led him to dismantle and reassemble several models at Killingworth Colliery, to discover how they worked, and he eventually became the colliery's engine-wright. To earn extra cash he was also a watch repairer, and he devised a new safety lamp for miners as well.

By 1814 he had built his first locomotive and soon he persuaded pit managers to let him build an eight-mile track from Hetton Colliery to the coast, to transport coal. He worked with his brother, also called Robert, on this project, and when it was opened in 1822 locomotives pulling eight wagons at a time, each holding more than two and a half tons of coal, used the line along its flatter portions. The Hetton railway ran uphill and, for these steeper inclines, a stationary engine was used to move the coal. For the first time in pit history no animals were needed for hauling coal.

This early line convinced George Stephenson of the need for cuttings and tunnels, to keep a line as flat as possible. With future projects father and son generally collaborated to bring their work to fruition. In fact, there's speculation that George depended on the expertise of his son to pull off the most ambitious projects attached to his name. Certainly, George was more of a broad-brushstroke individual, whose time cultivating rich and powerful acquaintances had as much to do with his success as the nuts and bolts of the steam engine. By the end of his career he was perhaps more polished at marketing than mechanics, an entrepreneur with interests in railway companies, locomotive manufacturers and coal, iron and lime mines.

Robert Stephenson (1803–1859).

At the insistence of his father, Robert Stephenson had enjoyed the benefits of an education. But his first job in 1819 was as an apprentice in a colliery and he was always aware of the plight of ordinary people, even while he rubbed shoulders with the wealthy and powerful. His reputation was for even-handed treatment of fellow engineers and navvies alike. 'The true and full effect of railways would not take place,' he insisted 'until they were made so cheap in their fares that a poor man could not afford to walk.'

He joined his father George in surveying the proposed route for the Stockton to Darlington railway after it was authorised in 1821. Following a spell at Edinburgh University, he joined the team building it. On its opening in 1825, the Stockton & Darlington was the first purpose-built locomotive-driven railway in Britain capable of carrying passengers and freight. It was from the outset a highly successful venture, meeting local needs and running at a profit for investors.

In 1824 Robert's health declined and he went to live abroad for three years, spending time establishing a rail network in South America. When he returned to the UK there were numerous projects that clamoured for his attention. After the success of the Stockton & Darlington Railway the indications were all good for a link between Manchester and Liverpool. Already plans were in place for this, although belligerent landowners and protective canal operators did their best to stymie the scheme. The necessary parliamentary bill was passed in 1826 and construction of the tracks, the bridges and viaducts needed for the 35-mile route got underway. All that was then needed was to persuade the board of the Liverpool & Manchester Railway that locomotives were the better option for pulling trains at a time when

The opening of the Stockton & Darlington Railway, 27 September 1825.

the most obvious choice still seemed to be horses, although stationary engines using hauling equipment were also closely considered.

To nail the debate, trials between the different options were organized at Rainhill in 1829 and the *Rocket*, built by George and Robert Stephenson, was triumphant after achieving an impressive speed of 30 mph. There were other engineers competing that day – Timothy Hackworth, John Braithwaite, John Ericsson – but history is written by the winners and other locomotive builders from that day to this were eclipsed by the success of the Stephensons.

When the line was opened a year later, their glory was marred by the death of William Huskisson, at the time the most prominent victim of a railway accident. He was the local MP and a former president of the Board of Trade when he was mowed down by the *Rocket*, with engineer Josiah Locke at the wheel. Sympathy has to go to Huskisson, who wandered on to the tracks at a time when the danger from trains was barely broadcast.

Robert Stephenson's first major solo project was the London to Birmingham railway, which received the necessary go-ahead in September 1833. Although there were some serious difficulties in its construction, not least building a tunnel at Kilsby, south of Rugby, after the good folks of Northampton resisted the railway's arrival. (The tunnel also fitted better with Stephenson's policy of flat lines as there was an incline at Northampton.) The line was opened in stages from 1837. The first train to travel between London and Birmingham on 17 September 1838 took four and a half hours. At its Birmingham terminus, Curzon Street, a passenger could link to the Grand Junction Railway and head onward to Manchester and Liverpool.

THE FIRST TRAIN TO TRAVEL BETWEEN LONDON AND BIRMINGHAM ON 17 SEPTEMBER 1838 TOOK FOUR AND A HALF HOURS

After that, Stephenson's services were in demand across Britain and the world. A keen yachtsman, Robert had little time to bob about on the waves thanks to the weight of work. At one point his name was associated with 160 different projects from 60 separate railway companies. He became a Conservative MP, president of the Institution of Mechanical Engineers for five years and later president of the Institution of Civil Engineers. He was showered with honours from countries across the globe but declined a knighthood.

Although he often spoke and acted in opposition to Brunel, the two men were friends. This friendship, like his one with engineer George Bidder, was rooted in a mutual fascination for mechanical ingenuity and a sense of being in the first wave of railway engineers who would leave an indelible mark on Britain.

Speaking to friends in Newcastle in 1850, Robert declared:

It seems to me but as yesterday that I was engaged as an assistant in laying out the Stockton and Darlington Railway. Since then, the Liverpool and Manchester, and a hundred other great works have sprung into existence. As I look back upon these stupendous undertakings, accomplished in so short a time, it seems as though we had realized in our generation the fabled powers of the magician's wand. Hills have been cut down and valleys filled up; and when these simple expedients have not sufficed, high and magnificent viaducts have been raised, and, if mountains stood in the way, tunnels of unexampled magnitude have pierced them through, bearing their triumphant attestation to the indomitable energy of the nation, and the unrivalled skill of our artisans.

On his death Queen Victoria gave permission for the funeral cortège to go through Hyde Park on its way to Westminster Abbey, where he was

A ventilation shaft on the Kilsby Tunnel, designed by Robert Stephenson.

buried, and the route was lined with thousands of mourners keen to pay respects to a man who had done so much to change the landscape.

It's only possible to imagine the sense of awe that might have filled the fishermen of the Tweed who watched as the Royal Border Bridge rose brick by brick to the lofty heights of its eventual splendour. *Bradshaw's* gives some insight into what life was like in Berwick 150 years ago, when the railways took salmon produced in the vicinity to London, packed in ice.

> **The salmon fisheries in the Tweed, once worth about £15,000 a year, have declined to £4,000. About Christmas the people here eat kippered salmon and plum pudding ... Much whisky is exported.**

Charting the journey to the next stop, Alnwick, *Bradshaw's* refers to 'that courageous heroine Grace Darling', an icon of the Victorian age lauded for her courage in helping to rescue passengers from a shipwreck off the Farne Islands.

Continuing down the line, using the Morpeth junction it was possible to join the Newcastle & Carlisle Railway, the first east-to-west railway link, completed in 1839. A branch line from it leading to Alston was in business from 1852. It was on the Newcastle & Carlisle Railway that a small but enormously significant leap was taken in railway administration, thanks to the foresight of one man.

For some time railway tickets took the same shape as stagecoach tickets had before them. They were nothing more than scraps of paper, only sometimes bearing scrawled details of the journey purchased.

Two tickets issued by the Great Western Railway (GWR) in 1864 and 1865.

(24) GREAT WESTERN RAILWAY.
No. 71 Feb 9 1864
PASS Mr. Davies by o'Clock Train,
from Wolverhampton to Oxford
Why granted
Specially engaged on business of the Company.
Available for One Journey only.
1 Class.
N.B. This Ticket is required to be shown to the Attendant on taking your seat, and delivered to the Collector at the end of the journey.
UP.

(24) GREAT WESTERN RAILWAY.
No. 52 Nov 19th 1864
PASS Mr. Davies by o'Clock Train,
from Swindon to Wolverhampton
Why granted Mr Armstrong's requisition
Specially engaged on business of the Company.
AVAILABLE FOR ONE JOURNEY ONLY.
1st Class.
N.B. This Ticket is required to be shown to the Attendant on taking your seat, and delivered to the Collector at the end of the journey.
UP.

GRACE DARLING (1815–1842)

Grace Darling was the seventh child of the Longstone lighthouse keeper, who spotted the shadowy outline of the four-year-old steamship *Forfarshire* before dawn on 7 September 1838. The ship had been sailing between Hull and Dundee with a cargo of cloth, soap and engineering equipment as well as 60 passengers and crew when it was driven onto rocks. A dozen survivors were left clinging to a rock off the Farne Islands greasy with pounding waves after the ship broke in two.

Grace Darling's heroic rescue of 7 September 1838.

Recognising their plight, 22-year-old Grace and her father William oared their own open boat a mile through rough seas to reach the rock, where a woman, her two children and nine men were stranded. Although they were still being gripped by their mother, the children were already dead. Grace and her father persuaded the woman to return in the boat with an injured man and three others. William, helped by two survivors, then set off a second time to rescue the remaining survivors. The bodies of the two children and a man were recovered by the Sunderland lifeboat later that day.

After word of the daring rescue spread, Grace Darling's life was turned upside-down as she became the object of national adoration. Artists painted her portrait; poets, including William Wordsworth, dedicated verses to her. There were Grace Darling chocolates produced by Cadbury's, and she was even offered a role in a circus.

At Longstone lighthouse letters containing high praise, gifts and money – including a £50 donation from Queen Victoria – turned up daily. More significantly, she and her father received gold medals from the Royal Humane Society and silver gallantry medals from the National Institution for the Preservation of Life from Shipwreck, the forerunner of the Royal National Lifeboat Institution. (At the time only gentlemen received gold medals. Those of lesser birth received reduced honours and sometimes only money.)

Although she didn't relish the public spotlight, she only had to endure it for a short time. Grace died of tuberculosis in 1842.

Thomas Edmondson ushered in a new age when he invented a railway ticket that could be punched by a station guard and bore a serial number. Before joining the Newcastle & Carlisle Railway in 1836 as stationmaster at Milton (later called Brompton) Edmondson had been a cabinet maker, and he still possessed some of the tools of his trade. In spare moments he began to make machinery that could produce cardboard tickets for train trips. His first efforts were comparatively rudimentary. The ticket number had to be filled in by hand and the mechanics of the validation stamp left railway staff in fear for their fingers. Still, Edmondson could clearly see a future for better ticketing.

When he received scant support from railway management, he changed jobs and began working for the Manchester & Leeds Railway Company when it opened in 1839 with an enhanced salary as chief booking clerk. However, he still sought to perfect the apparatus so it could produce fully printed tickets with ease. After he teamed up with mechanical engineer William Muir, Edmondson finally left his job at the railways and achieved his ambition. He patented the design – no mean feat for those days when the system was weighted with bureaucracy – and charged railway companies 10 shillings per year per mile for its use. His invention and its application to the new fashion of train travel made him a rich man.

IN 1844 JUST 805 MILES OF LINE WERE GIVEN GOVERNMENT APPROVAL. THE FOLLOWING YEAR THE FIGURE WAS 2,700 MILES

The railways didn't make everybody rich, though. In the mid-1840s a rampant railway mania gripped the country and ended with many people suffering grievous financial losses. It was more than simple economics that led to ruin for many.

Although new railway lines needed parliamentary approval they were built by private companies that had to raise funds for the necessary building work. At a glance, the provision of railways seemed like a sure-fire way of making money, and railway schemes became a popular choice for investors. After all, railways were welcomed as a swift alternative to uncomfortable and sometimes dangerous coach travel. They enabled existing and new businesses to function. Despite a slow start, railways had won the hearts and minds of the British public by the 1840s and financial advisors of the day were impressed enough to funnel all the customers they had into railway investment. For the first time, entrepreneurs who alighted on railways as a means to make

Railway speculators in Capel Court, a lane leading to the Stock Exchange where dealers congregated to do business.

money could not only find investors among the aristocracy but also in the burgeoning ranks of the middle classes. But there was no national strategy in place to create a sound, sensible railway system, nor were there safeguards at hand if things went wrong.

For a while everything went according to the railway companies' plans. Nationally the economy appeared to be booming, and rates of interest were favourable to investors. Lots of companies offered enticing deals. People only had to produce 10 per cent of some investments in hard cash while the rest was 'on a promise'.

Meanwhile, the business of railway building was indisputably expanding at an immense rate. In 1844 just 805 miles of line were given government approval. The following year the figure was 2,700 miles. At the planning end of countless schemes, Brunel was something of a lone voice when he made his objections to railway mania known in a letter:

> *I am really sick of hearing proposals made. I wish it were at an end. I prefer engineering very much to projecting, of which I keep as clear as I can … I wish I could suggest a plan that would greatly diminish the number of projects; it would suit my interests and those of my clients perfectly if all railways were stopped for several years to come.*

The Scottish historian Thomas Carlyle (1795–1881).

Generally, the interest of the public was not waning; indeed, it was sharpened by the prospect of making quick, clean money. Most investors, however, failed to ask the right questions. There were schemes for railways that went to places no one wanted to go. Others were planned for inhospitable terrain that would be costly to bridge. Many plans were poorly costed, and in numerous cases there was sharp practice among some businessmen who had an eye for a fast buck. In general, the promises made in newspaper advertisements about the proposed benefits of railway investment were at best optimistic and at worse downright fraudulent.

Landowners didn't help, either. Believing railway construction to be a golden goose, they demanded high prices before giving up land for track beds. Railway companies were also obliged by law to fence every foot of line they laid. As a consequence it was much more expensive to build railways in Britain than anywhere else in the world.

Inevitably, with so much of the promised capital amounting to no more than a fiction, there was an economic implosion that left numerous first-time investors out of pocket. The historian and commentator Thomas Carlyle explained it like this:

> *Then railways bubbled. New ones were advertised, fifty a month, and all went to a premium. High and low scrambled for the shares, even when the projected line was to run from the town of Nought to the village of Nothing across a goose common. The flame spread, fanned by prospectus and advertisement, two mines of glowing fiction, compared with which the legitimate article is a mere tissue of understatements.*

There was one final, optimistic rush in 1845, as solicitors sought to bring new plans to their local clerk of the peace by 30 November, so they could be included in the next parliamentary session. No fewer than 879 plans missed this deadline, causing considerable panic among their investors. But it was worse still for those who gained approval and began spending money that was promised, fraudulently or otherwise, only to find the company purse all but empty.

As the juggernaut driven by railway investment juddered to a halt after 1846 there were some benefits to the long-term security of train travel. Small companies that failed in the face of unchecked business chicanery were amalgamated with larger ones, making better sense of long-distance railway journeys for the train-travelling public. One early example of this was the Midland Railway, formed in May 1844 out of numerous lines proliferating around Birmingham and Derby. In 1848 the London & Birmingham, Grand Junction and Liverpool & Manchester companies came together to form the London & North Western Railway. One useful spin-off that followed was a limit on the number of stations. Instead of towns getting numerous railway stations to serve different companies – at one stage Swansea boasted no fewer than six termini – just one or two were built.

George Hudson invested his inheritance in railways and eventually controlled over 1,000 miles of track.

But there were a string of disaster stories, with many people left destitute. Those who borrowed to furnish railway companies with an initial stake were particularly hard hit. John Francis, who wrote two volumes on *The History of the English Railway*, published in 1851, was emotional on the topic: 'It reached every hearth, it saddened every heart in the metropolis. Entire families were ruined. There was scarcely an important town in England but what beheld some wretched suicide. Daughters delicately nurtured went out to seek their bread. Sons were recalled from academies. Households were separated . . .'

George Hudson was one of Britain's prominent businessmen whose bloated speculation lit the fuse of railway mania. When the bubble popped he was left in chronic debt and shamed by his risky business practices. However, in Sunderland, where he was the MP, York and Whitby he was held in an unshakeable affection, having brought the railway and all its advantages to those places.

Hudson was the son of a farmer, born just outside York at the turn of the nineteenth century. He moved to York to work in a drapery business, becoming a partner after marrying the daughter of his boss. His ambition heightened considerably with an inheritance of some £30,000 from a distant relative in 1827. He indulged himself in the two abiding passions of his life: railways and politics. During his tenure as Lord Mayor of York he invested in the North Midland Railway.

MIDLAND RAILWAY.
Adelphi Hotel, Liverpool
FORTH BRIDGE
Midland Rᵗ Cᵒ Queen's Hotel, Leeds
GREENOCK PIER
GLASGOW
EDINBURGH
CARLISLE
NEWCASTLE
LANCASTER
YORK
LEEDS
BRADFORD
LIVERPOOL
MANCHESTER
SHEFFIELD
LINCOLN
DERBY
NOTTINGHAM
LEICESTER
PETERBORO'
BIRMINGHAM
WORCESTER
HEREFORD
BRECON
CHELTENHAM
GLOUCESTER
BEDFORD
CAMBRIDGE
SWANSEA
CARDIFF
BRISTOL
BATH
ST PANCRAS
LONDON
Matlock
BATH
Midland Hotel, Bradford
Bournemouth
Midland Grand Hotel, St Pancras, LONDON.
BEMROSE & SONS, PRINTERS, DERBY & LONDON.
GEO. H. TURNER, GENERAL MANAGER, DERBY.

A poster produced for the Midland Railway to promote their rail routes, c. 1900.

He saw to it that a planned route for London dropped by his home city, and the profits rolled in with ease. In 1833 he formed his own railway company, the York & North Midland Railway, to link other Yorkshire populations. He raised some £446,000 in capital and the line was finished by the middle of 1839.

His next aim was to lay a line between York and Newcastle that would, he realised, be an important part of an east coast line to link London with Scotland. To win official permission for the work to take place, he distributed some £3,000 in bribes.

By 1844, through his various railway companies, he controlled an estimated 1,000 miles of track, and the chaos that was engulfing some railway companies seemed to be passing him by. Observers tagged him 'the Railway King' and for a while it must have seemed his star would never fade. He became Conservative MP for Sunderland in the 1845 election and used his newly won position to fight against a proposal by William Gladstone for partial state ownership of railways, as the first victims of railway mania began falling by the wayside. If he wasn't enhancing his own business, he spent time sabotaging the chances of rivals.

One of his close friends was the Duke of Wellington, both hero and villain to the British public and a man who wielded considerable power in and out of Parliament. Wellington had become Prime Minister in 1828 but painfully misunderstood the political process (After his first cabinet meeting he remarked: 'An extraordinary affair. I gave them their orders and they wanted to stay and discuss them.') Wellington – who ring-fenced his home against railway development – harboured reservations about the railways after witnessing the death of William Huskisson at the opening of the Liverpool & Manchester Railway in 1830. He was also concerned that the railways offered travel opportunities to ordinary people – something which earned his hearty disapproval.

TO WIN OFFICIAL PERMISSION FOR THE WORK TO TAKE PLACE, HUDSON DISTRIBUTED SOME £3,000 IN BRIBES

As late as January 1846 the *Standard* newspaper lauded Hudson's work. 'Two hundred thousand well-paid labourers, representing as heads of families, nearly one million men, women and children, all feast through the bold enterprise of one man. Let us hear what man or class of man ever before did so much for the population of a country.'

But it was at about this time that his friendship with railway godfather George Stephenson petered out, with the engineer concerned about many of Hudson's business methods. And Stephenson's suspicions were justified. As the shakedown sparked by rail mania continued Hudson found himself in difficulty.

It was a desire to spike the guns of a rival business that made him take over the Eastern Counties Railway in 1846. To do so, Hudson trebled the dividends of shareholders, illegally dipping into company capital to do so. In short, he was guilty of embezzlement. Although it was a widely used device in business it drew unwelcome scrutiny to his business dealings. Soon the fact that he used inside information to massage share prices, that he'd lied to shareholders about the viability of his various businesses, and that he had sold land he didn't own to the Newcastle & Berwick Railway came to light.

The Times reflected public enmity towards Hudson and his role in the railway mania in 1848, saying:

> *It was a system without rules, without order, without even a definite morality. Mr Hudson, having a faculty for amalgamation and being so successful, found himself in the enjoyment of a great railway despotism, in which he had to do everything out of his own head and among lesser problems to discover the ethics of railway speculation and management.*

Hudson was compelled to resign as chairman of all the railway companies in his portfolio. Although he remained an MP until 1859 he could not shake off the odour of corruption, nor would he pay back shareholders' money they had lost through his dubious dealings. As a result he was imprisoned in York Castle in 1865 for debt. His friends clubbed together to settle his liabilities and as a result he was behind bars for only 16 months. Five years later he died, having both fuelled railway mania and fallen foul of it.

It wasn't the end of railway mania, which returned to haunt industrialists including Sir Samuel Morton Peto, who built widely in eastern England, in 1866. By this time, though, the network of lines around the UK was looking more like a finished article than ever before, with major cities linked one to another – although, incredibly, it was set to double from its 1860 size before railway construction came to an end. (One enduring argument put forward by business rivals against the construction of the Great Central Line, which came

Bradshaw's railway map of Great Britain and Ireland, 1901, showing the huge expansion in railway construction from 1860 onwards.

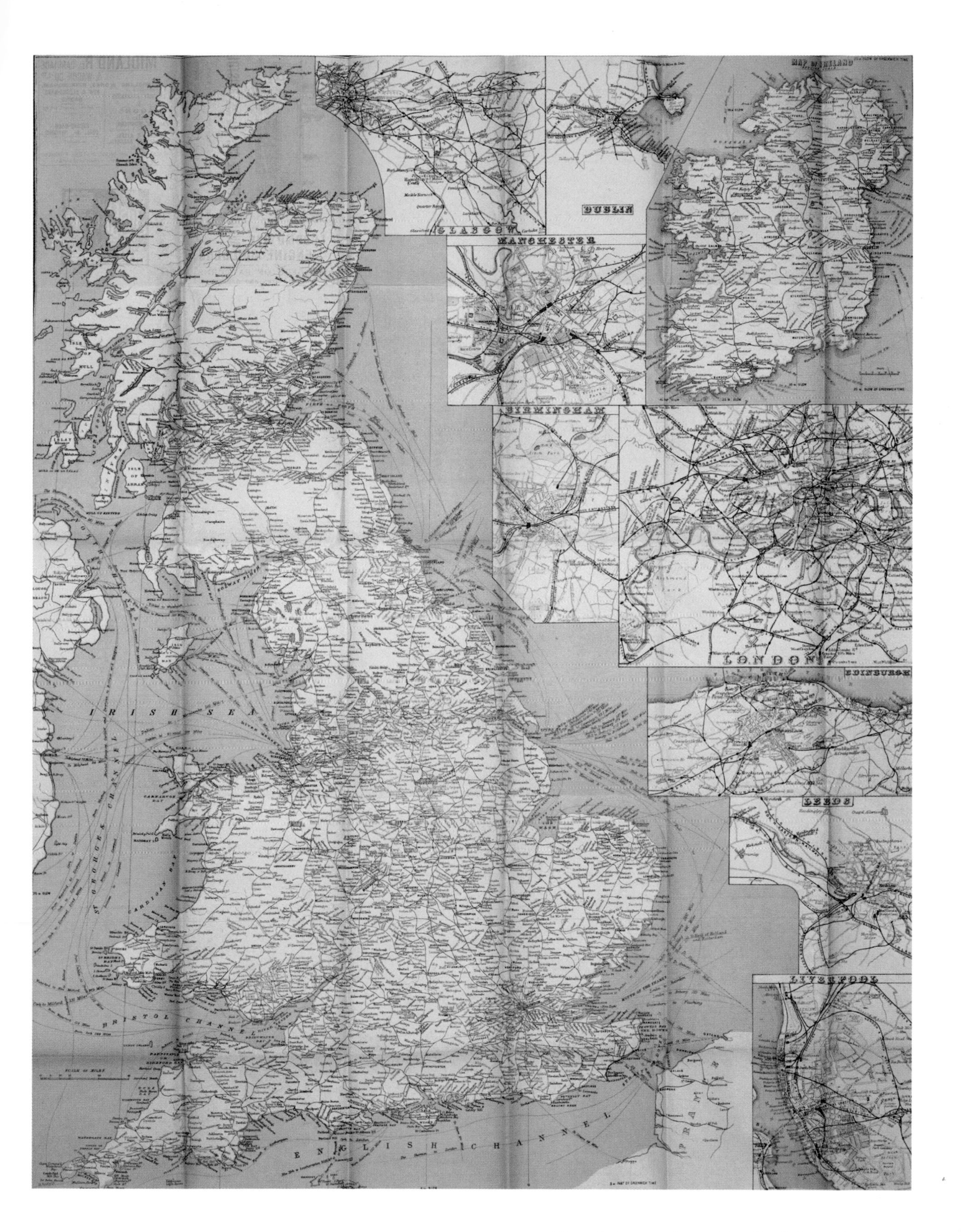
GLASGOW
DUBLIN
MAP OF IRELAND
MANCHESTER
BIRMINGHAM
LONDON
EDINBURGH
LEEDS
LIVERPOOL
IRISH SEA
BRISTOL CHANNEL
ENGLISH CHANNEL

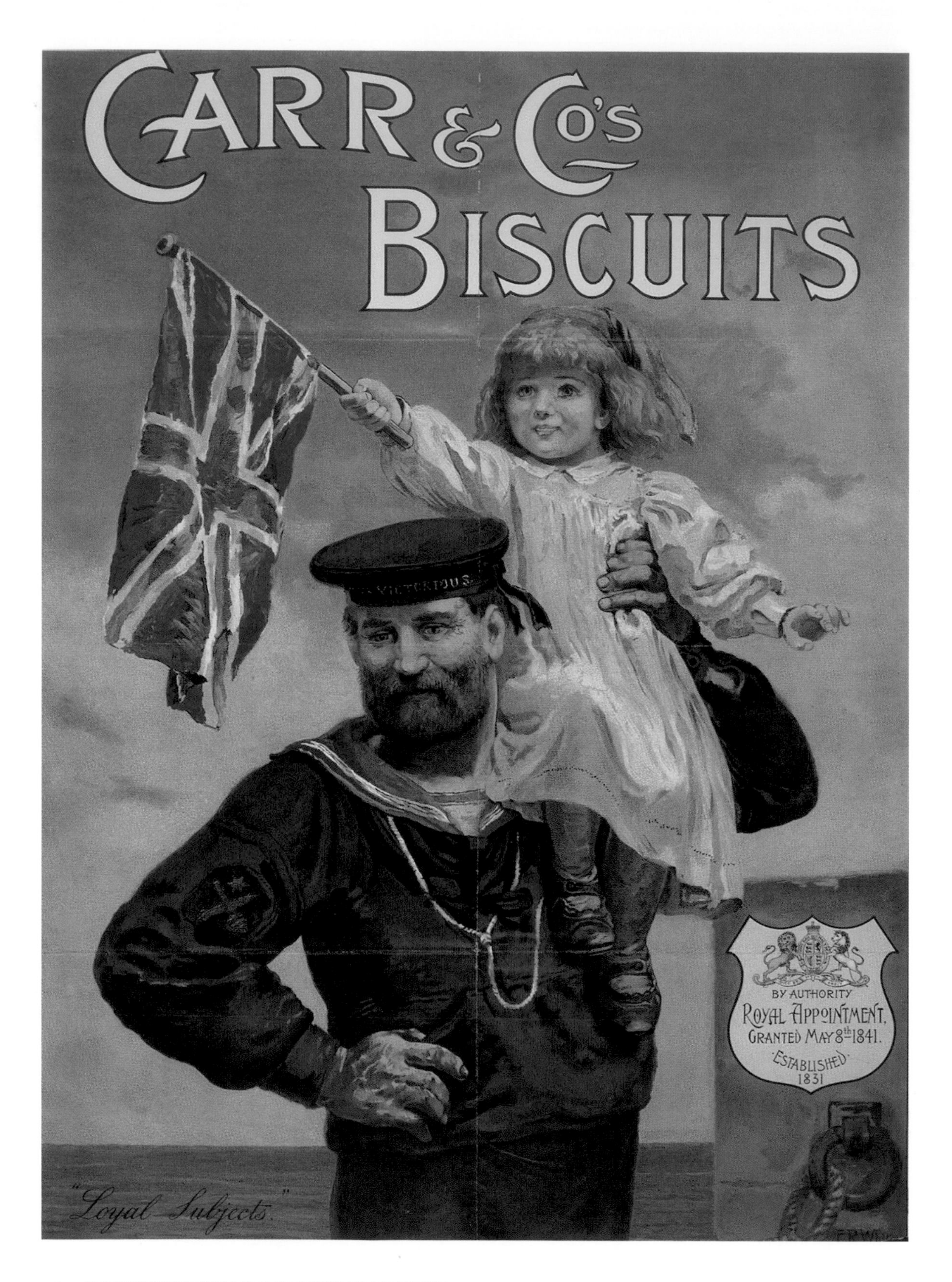
CARR & Co's
BISCUITS
BY AUTHORITY
ROYAL APPOINTMENT.
GRANTED MAY 8th 1841.
ESTABLISHED
1831
"Loyal Subjects"

NO FEWER THAN SEVEN DIFFERENT COMPANIES RAN THEIR ENGINES INTO CARLISLE'S CITADEL STATION

into the newly built Marylebone Station in London via Nottingham, Leicester and Rugby in the 1890s, was that there were plenty of rail services already doing the same job.)

So much for the north east, where early episodes of railway history were wrought. Across the neck of the country in a landscape marked by Hadrian's Wall, the railway ploughed through heather and bracken to link two centres of population. At the Carlisle end of the Newcastle to Carlisle line there is, concedes *Bradshaw's*, a cathedral, a port and an eight-sided brick chimney 305 feet high. But the guide's most vivid description is for the biscuit factory Carr & Co, opened by Jonathan Dodgson Carr in 1837, the year Victoria came to the throne. The bread and biscuits Carr's made were so well regarded the firm received a royal warrant from the Queen in 1842. *Bradshaw's* concurs, saying:

> **[Carlisle] is also celebrated for its manufacture of fancy biscuits, which are produced in a most complete state, all by machinery and to an extent that would certainly astonish visitors ... if curiosity should induce the tourist to make a visit to the manufactory of this noted firm, we do not hesitate to say that it would be found highly interesting. If any prejudice exist against the free use of fancy biscuits, it will at once be removed on an inspection of the works and the process of production, even from the minds of the most fastidious – the most scrupulous cleanliness being observable throughout the whole works.**

What the guide fails to note is the extensive railway workings in the city. At one time no fewer than seven different companies ran their engines into Carlisle's Citadel Station, and its marshalling yard was one of the biggest in Europe.

The next stage of the journey runs from Carlisle down to Barrow-in-Furness. The rump of land beneath Barrow was kept remote from the rest of the country, not least because of the racing waters that filled Morecambe Bay to the south. But railway accessibility from Carlisle

LEFT: **A poster advertising Carr's biscuits.**

OVERLEAF: **The iron and steel works at Barrow c. 1880. A print from *Great Industries of Great Britain Volume I*, published by Cassell, Petter and Galpin c. 1880.**

changed all that. Thanks to Victorian technology this previously isolated region became the home of cutting-edge industry. In the early decades of the eighteenth century there was small-scale trade in iron ore, richly deposited around the region.

The Furness Railway, distinguished by marine blue and white livery, was originally a secluded line built in 1846 in response to this limited industry. Locomotive no. 3, built in 1843 and called *Old Coppernob*, used to ply this route. Yet the Furness Railway emerged as the most famous line in the area, although the coast-hugging Maryport & Carlisle Railway was completed by 1845 while the Whitehaven Junction Railway was in operation two years later. Also in 1847, a spur was laid inland from Workington to form the short-lived Cockermouth & Workington Railway. In 1865 the tracks were re-aligned and a new station built for the Cockermouth, Keswick & Penrith Railway.

A relentless campaign to keep the railways out of the Lake District hindered progress but did not stop railway expansion. Opposition was

From 1907 to 1941 the famous Furness Railway steam locomotive *Old Coppernob* was preserved in a special glass case outside the station – it is now in the National Railway Museum at York.

largely founded in a belief that ordinary people would not be able to appreciate the beauty of the lakes and would probably wilfully ruin the place. Poet Laureate William Wordsworth was a key figure in the anti-train lobby before his death in 1850. When a railway link between Kendal and Windermere was proposed he wrote a poem opposing it, with the first line: 'Is then no nook of English ground secure from rash assault?' Although construction was postponed the line was eventually built, albeit with a modified route.

A later pamphlet directed at railway companies addressed the perceived difficulties of bringing working people to the Lake District.

> *...can't you teach them to save enough out of their year's wages to pay for a chaise and pony for a day, to drive Missis and the Baby that pleasant twenty miles, stopping when they like, to unpack the basket on a mossy bank? If they can't enjoy the scenery that way, they can't any way; and all that your railroad company can do for them is only to open taverns and skittle grounds round Grasmere, which will soon, then, be nothing but a pool of drainage, with a beach of broken gingerbeer bottles; and their minds will be no more improved by contemplating the scenery of such a lake than of Blackpool.*

Writer John Ruskin – whose father-in-law was a victim of railway mania – took up the baton from Wordsworth and declared:

> *I have said I take no selfish interest in this resistance to the railroad. But I do take an unselfish one. It is precisely because I passionately wish to improve the minds of the populace, and because I am spending my own mind, strength, and fortune, wholly on that object, that I don't want to let them see Helvellyn while they are drunk.*

Indeed, the Buttermere & Braithwaite line, proposed in 1882, fell through after orchestrated opposition. But some tendrils of tracks made their way into the Lake District from Furness Railway territory. Mostly the trains helped local communities make the most of their natural resources. Ruskin's ire for railways was raised when the Furness Railway built a line in close proximity to Furness Abbey in 1864.

For a description of the industry in 'prosperous' Whitehaven in 1866, there's no need to look further than *Bradshaw's*:

POET LAUREATE WILLIAM WORDSWORTH WAS A KEY FIGURE IN THE ANTI-TRAIN LOBBY BEFORE HIS DEATH IN 1850

> **The coal measures form a thin strip round the coast past Workington and Maryport. The mines are worked by deep shafts a quarter of a mile down, close to the edge of the sea, under which they run more than two miles, the dip of the beds being as much as one ft in ten. Some of them are eight and ten ft thick with good coal. When raised the 'black diamonds' are turned into the wagons which descend the tram road by their own weight to the quay and drag up the empty ones; here they are dropped through the wooden lurries into the vessel's hold.**

A Furness Railway No. 20 steam engine at the recreation of Rowley Station, Beamish Museum, County Durham.

The Furness Railway had many success stories but one failure. It seemed to be the perfect way to bring tourists to Seascale, a small coastal settlement that seemed packed with potential. The railway's general manager, James Ramsden, earmarked the place for development as a resort in 1879, believing it a perfect location from which to explore the Lake District, and he encouraged a Liverpool architect, Edward Kemp, to draw up plans for a hotel, villas and seaside promenades that would befit a prestigious Victorian holiday destination. Although some notable dwellings were built, the momentum faltered and Seascale never rivalled places like Southport or Scarborough.

Bradshaw's was sensitive to the changing landscape in the area and notes of Furness that: 'Iron is now forged in this vicinity where the stag, wolf and wild boar were formerly hunted.' Industry hastened in after 1850 when speculator Henry Schneider discovered a large new source of iron ore. Applying the vision that distinguished many Victorian businessmen, he went about exploiting this resource using the railways.

He became one of the movers and shakers behind the railway, which not only served the numerous iron ore mines that were by now in operation in Lindal but also the blue slate mines at Kirby. The railway headed for the port at Roa Island, from where ships had access to other British ports and the rest of the world. There were a series of extensions and branch lines until the area was webbed with lines and Barrow, Whitehaven, Workington and Maryport became prominent harbours.

Nº 20

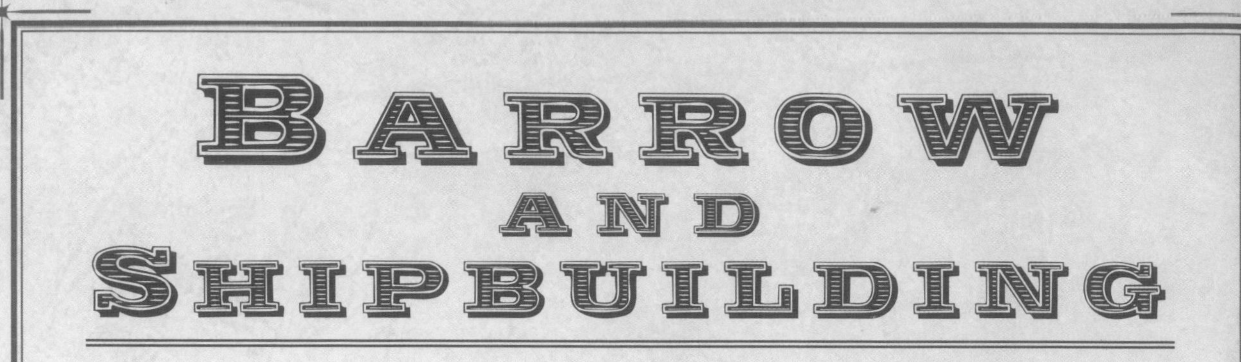

Barrow and Shipbuilding

In a natural progression, a new generation of industrialists sought to make ships in Barrow out of the steel that was readily at hand. (With opportunities abounding in the 1870s it was said that there were more aristocrats per head of population in Barrow than anywhere else in Britain.) Despite taking immense pride in the Royal Navy, Britain lagged behind the French as shipbuilding moved from the use of wood through iron cladding to steel.

For several decades the Royal Navy was content to assert itself by numerical superiority alone. However, the end of the nineteenth century saw an escalating arms race, primarily between Britain and Germany. Britain could no longer rely on having plenty of ships to win supremacy at sea if they weren't modern and powerful.

Barrow began building naval ships at the behest of the Admiralty in 1877, and within a decade had branched out into submarine construction. But the Royal Navy was not the shipyard's only customer. The first submarines – built to the specification of Swedish industrialist and arms dealer Thorsten Nordenfelt – were sold to Turkey and Russia. By 1888, reflecting its international role, the shipyard was renamed the Naval Construction and Armaments Company.

A railway being used to feed the blast furnaces at the Barrow Haematite Iron and Steel Company, 1890.

By 1897 shipbuilding was a bigger employer than both the railway and the steelworks.

Still its expansion was not complete. The company was bought by Vickers in 1897, after which it claimed to be 'the only shipbuilder capable of designing, building, engineering and arming its own vessels'. Four years later it was the Barrow shipbuilders who got to work, in secrecy, on the first five submarines ordered by the Royal Navy, using a blueprint from Irish-American John P. Holland. The first submarine was consequently generally known as Holland 1. By 1897 shipbuilding was a bigger employer than both the railway and the steelworks.

The effect on the area in five short decades was immense. Midway through the nineteenth century the population of Barrow was a mere 700. By 1865 the number had grown to 10,000 and by 1881 it stood at 47,000.

The success of initial iron-based industrial ventures fired the imagination of Schneider and Ramsden. There was more profit to be had in smelting the iron ore on the peninsula in Barrow to make steel. British industrialist Henry Bessemer discovered a cheap way to convert pig iron into steel and established steel works in his home town of Sheffield. When the patent restrictions on his method ran out in 1870 it was adopted by others, including Schneider and Ramsden. Consequently the largest steel works in the world sprang up in this somewhat unlikely corner of the country.

There is a thought-provoking footnote linked to the railway lines around the Furness peninsula. On 22 September 1892 a steam engine towing a goods train was stopped just beyond Lindal passenger station when driver Thomas Postlethwaite felt the ground beneath him start to shake. He leapt off before the engine lurched forward as the ground beneath the tracks began to yawn. Fortunately no one else was in the immediate area.

A gang of recovery workers soon uncoupled the wagons and they were pushed back to the safety of the station. Retrieving the 35-ton engine was a more difficult prospect. The best plan seemed to be to dig away an incline on which tracks could be laid, then the engine could be hauled to safety. But as the digging began the earth rumbled again, this time swallowing the engine entirely. No one knows how far into the Parkside mine workings beneath the railway the engine sank, although it was never seen again after its final descent. Inside the cab Postlethwaite's jacket containing his long-service gold watch remains to this day.

The railway shadowed the coastline around Morecambe Bay, striding across two significant stretches of water, until it reached Carnforth. The line between Carnforth and Whitehaven was opened in 1857, offering links to the Lake District via the Kent Viaduct. When it was built in 1857 a new technique was employed by engineer Sir James Brunlees. The supports were tubular cast-iron piles with large discs at their bases, jetted into place through the shifting sands of the River Kent and filled with concrete. A similar viaduct was built on the same lines at Leven. The single-track Kent Viaduct was completed at a cost of little more than £15,000. In 1863 the Furness Railway Company

OVERLEAF: A view from Arnside Knott over the Kent Viaduct to Whitbarrow and the Lakeland fells.

widened the viaduct so it could accommodate twin tracks. However, its ironwork deteriorated with comparative speed and by 1915 the weakened piles were encased in concrete and brickwork. Engineer Brunlees was also behind the Mersey railway and piers at Southport, Llandudno and Southend.

Initially, Carnforth was a small station opened in 1846 on the Lancaster & Carlisle line. A decade later it was transformed into a busy junction when the Furness and Midland tracks met there. The station was sufficiently important to be rebuilt entirely in 1880 at a cost of £40,000. And it was at this new station that Prime Minister William Gladstone was publicly humbled by Queen Victoria, to his intense indignation. Gladstone had been staying at Holker Hall with the Duke of Devonshire in February 1885 when news broke about the fall of Khartoum.

Sudan was a thorny subject for government and monarch. Gladstone believed it was an unnecessary conflict at the time, while

Carnforth Railway Station was later used as a location for the film *Brief Encounter*.

staunch imperialists – like the Queen – believed control of overseas dominions was pivotal. General Gordon had been dispatched to Khartoum to organise a withdrawal of forces. Instead he upped the defences and prepared for a siege. Vastly outnumbered by desert fighters, Khartoum was eventually breached and the defenders slaughtered. Against the specific orders of their leader, triumphant warriors paraded Gordon's head on a pole. The relief force eventually sent by Gladstone arrived two days later.

As Gladstone arrived at Carnforth for a hurried journey back to London, the stationmaster handed him a stinging telegram from Queen Victoria: 'These [sic] news from Khartoum are frightful and to think that all this might have been prevented and many precious lives saved by earlier action is too fearful.'

Even to the unpolitical eye of the stationmaster, the Queen's fury was evident. But Gladstone was also cross as the proper code that normally accompanied official missives like this had been dropped, and the split between Prime Minister and Queen was now obvious to all. Both the Queen and Gordon were popular figures. The débâcle at Khartoum was one of the major reasons that Gladstone lost the election later that year.

In railway terms, Carnforth was always overshadowed by its southerly neighbour, Lancaster, a historic city once isolated by the coast in the west and miles of inhospitable terrain to the east. Lancaster became a regional centre for justice and punishment.

The railway ran from Lancaster to Heysham, where the ferry to the Isle of Man set sail in the summer months. In the 1850s this rocky, self-governed island set between Scotland, England and Ireland became popular with tourists, not least thanks to pioneering travel agent Thomas Cook.

It was once claimed that Thomas Cook & Son ranked with the Roman Catholic Church and the Prussian Army in terms of efficiency in the Victorian age. His extraordinary powers of administration were matched by a fervent desire to see travel available to all. 'Railway travelling is travelling for the millions,' he declared in 1854. For him, trains breached the class divide as never before, with passengers including 'a mourning countess and a marriage party – a weeping widow and a laughing bride – a gray head and an infant of days'.

'RAILWAY TRAVELLING IS TRAVELLING FOR THE MILLIONS'

A portrait of Thomas Cook (1808–1892).

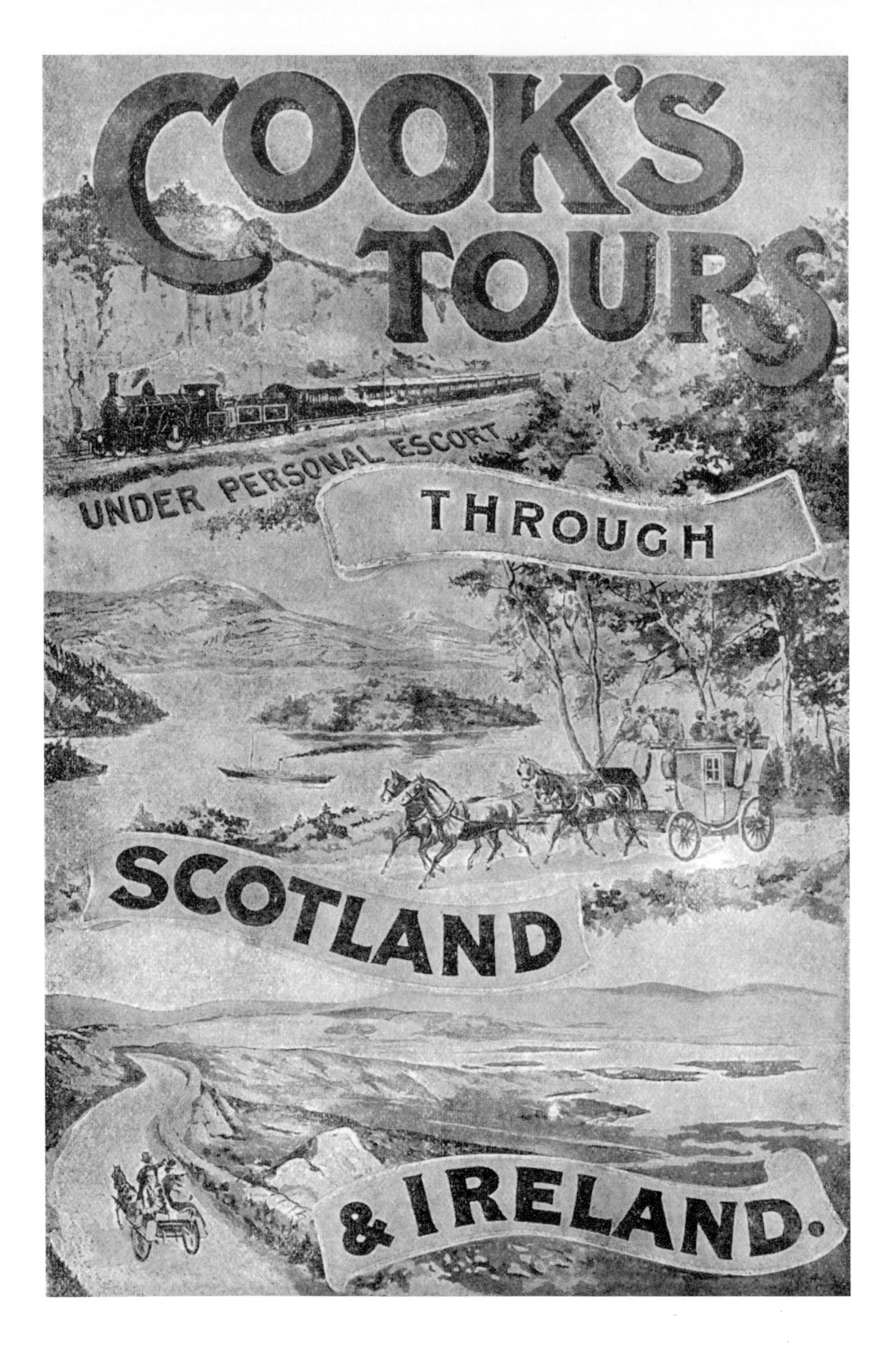
COOK'S
TOURS
UNDER PERSONAL ESCORT
THROUGH
SCOTLAND
& IRELAND.

'Cook's Tours through Scotland and Ireland' (c. 1900s)

He hit on the idea of chartering a train in 1841 on the 15-mile walk between his home in Harborough and the temperance meeting he attended in Leicester. After negotiations with the Midland Counties Railway Company he could offer cut-rate fares to fellow Temperance Society members for a scheduled meeting. Accordingly, 570 set off in third-class open carriages, or 'tubs' as he called them, between Leicester and Loughborough, to be met by a brass band.

Cook had fond memories of this first outing:

> *The people crowded the streets, filled windows, covered the housetops and cheered us all along the line, with the heartiest welcome. All went off in the best style and in perfect safety we returned to Leicester; and thus was struck the keynote of my excursions, and the social idea grew upon me.*

Initially he satisfied himself with organising this type of trip, until in the summer of 1845 he arranged an excursion to the seaside at Liverpool. Its administration wasn't easy as passengers would encounter four different railway companies en route. There were hotels and restaurants to consider. Nonetheless, after the trip was offered to the general public it proved so popular that all 350 tickets were gone within a week.

Cook's next target was Scotland, where large swathes of land were as yet untouched by track. Here he learned that steamer connections were as reliable as trains and equally as popular. However, it was London's Great Exhibition in 1851 that secured his future as a travel agent. There was grave concern among some of the exhibition organisers that thousands would be excluded from it as they lived too far away, could not afford the entrance fee or did not have the wherewithal to work out the journey.

They approached Thomas Cook with a view to organising cost-cutting excursions from the Midlands. For his part, Cook was delighted to play a role in peopling the Exhibition, which he thought would bringing a 'harmonising and ennobling influence', and he threw himself into finding cheap accommodation. He encouraged men in factories around the Midlands to begin Exhibition clubs, saving a small amount each week to cover the cost of the trip. He even launched a newspaper, *The Excursionist and Exhibition Advertiser*, to generate interest.

Victorian Crime and Punishment

The Victorian public had specific mores, and concern for law and order was among them. A fresh look at crime and criminals in the era inspired some radical changes in punishment, nowhere more apparent than at Lancaster.

The era of the Bloody Code – when a man could be hanged for any one of 200 different offences – was over. Transportation to Australia was also off the agenda after 1852, following complaints from Sydney about this booming country becoming a dumping ground for British villains. Instead came the age of incarceration. And before a prison-building programme could be got underway, Lancaster Castle, bleak and comfortless, filled the breach.

The castle's medieval cells were used as a men's jail from 1794. Thirty years later a women's prison was added to the castle, which followed the principles of the Panopticon advocated by Jeremy Bentham in the late eighteenth century. In it, single cells radiated out from the guards' post and were under constant surveillance. (It was a design that found greater favour in America than Britain.)

Until 1835 all major trials in the county were held at Lancaster and the court there was known as 'the hanging court' for giving the death sentence more often than anywhere else outside London. After that date Manchester became the busiest court and condemned prisoners were hung at Salford and Strangeways gaols.

But by Victorian times the number of capital sentences nationwide was falling, with only murderers and traitors given the death penalty. Between 1836 and 1840 the number of people hanged was just 10 per cent of that in Georgian times. Provincial executioners were no longer required, and a single hangman travelled across the country, using the burgeoning rail service to reach his destinations.

Only eight men were hanged at Lancaster during Victoria's reign, each to the sound of the mournful toll of the prison bell. The first of those was Richard Pedder on 3 September 1853, the last man to be hanged in Lancaster on a 'short drop', which usually resulted in a slow death by strangulation and was national hangman William Calcraft's favoured method.

Calcraft's successor, William Marwood, brought a measure of science to the business of killing by calculating the length of rope needed for a swift death by a broken neck. He also replaced the slip knot with a metal ring to speed the grisly event. Marwood was duly succeeded by James Berry in 1884, a religious man who became deeply opposed to capital punishment after becoming convinced he had killed innocent men.

Only occasionally did the hangman tread the platform at Lancaster station, opened in 1846. In 1862 Walker Moore cheated the hangman of a day's pay when he drowned himself in the prison cistern on the morning his execution was scheduled. Three years later, Stephen Burke became the last man to be publicly hanged at Lancaster. Five more men died at the State's behest in Lancaster gaol before the Edwardian age. The last man to die there was Thomas Rawcliffe, who was hanged in 1910.

Lancaster Castle.

Although hangings were less frequent in Victorian times, the daily drudgery for prisoners was numbing. In common with other prisons, Lancaster gaol used treadmills for those prisoners sentenced to hard labour. There were two inside the prison, one used to power calico looms that made material for prison uniforms and a second to draw water from a well.

The treadmill, also known as 'the shin scraper', was developed by Sir William Cubitt and was in use at Lancaster from 1822. Cubitt also helped to build the South Eastern and Great Northern railways and he was a key designer for the 1851 Great Exhibition. For the psychologically damning treadmill, however, he is less fondly remembered.

Prisoners trod the wheels for 10 hours a day, averaging 96 steps a minute. For five minutes in every 20 they were permitted to rest on a nearby stool. Between each prisoner on the drum of the wheel was a partition to prevent talking. The rule of silence in prisons at the time was held in high esteem by the Victorians as a way of giving convicts the time to reflect on their crime. Attempts at communication were swiftly punished. Religious services, held regularly in gaols, were enthusiastically attended and prisoners sang lustily, eager to hear the sound of their own voices and those of others.

Reformers, including Quaker Elizabeth Fry, were unsure about the new trend. She insisted: 'In some respects I think there is more cruelty in our gaols than I have ever before seen.' Much later, Charles Dickens wrote: 'I hold this slow and daily tampering with the mysteries of the brain to be immeasurably worse than any torture of the body.'

But the overriding concern was not rehabilitation but punishment. When Sir Edmund Du Cane took over responsibilities for prisons in 1865 he promised a concerned Victorian society 'hard bed, hard board and hard labour' for those behind bars. It would be years before free association was allowed between British prisoners.

An Isle of Man Railway steam train at Douglas Station.

By now other railway companies were catching on to Cook's ideas, and he found himself undercut by several. Undaunted, he matched their prices and vowed to learn some hard commercial lessons. Ultimately, an estimated 165,000 people visited the Exhibition on Cook's tours, some 3 per cent of the total number of visitors.

It was only a matter of a few years before Thomas Cook's business expanded to involve the Isle of Man. For the Victorians there was plenty to recommend the island, from its bracing climate to its stunning scenery. Bradshaw's guide gives some indication as to why people were drawn to the island. Barren and blowy, the slate and heath landscape is, the guidebook says, 'like a beggar's tattered coat'.

These healthy breezes, with the short, springy turf, reconcile the pedestrian to the wild, desolate character of the scenery only enlivened by a few small sheep and occasionally the skulking sheep-stealer. The view from the summit of the mountains embraces the island and the sea in which it is set, as far as the shores of England, Wales, Scotland and Ireland if the air is sufficiently clear.

There's a word of warning too for those so taken with the island that they are tempted to move there.

> **Strangers before becoming residents should make themselves well acquainted with the Manx laws, they being totally unlike those of England, Ireland, Wales or Scotland. Arrests for debt can be made even for a shilling on this island and execution follows instanter.**

UNDER MANX LAW, QUEEN VICTORIA WAS REQUIRED TO INDICATE WHETHER OR NOT SHE WISHED KEWISH'S SENTENCE TO BE COMMUTED TO LIFE IMPRISONMENT

Bradshaw's was right to point out that the Isle of Man enjoys its own jurisdiction. However, the implied threat that many were killed by capital punishment was mistaken. The last person to be hanged there was John Kewish Junior, who was hanged on 1 August 1872 after being convicted of murdering his father. Before that no one had died on a gallows there for three decades.

Kewish lived with his parents and sister at the time his father was found dead. Police believed a pitchfork was used as a weapon. At his first trial Kewish denied the charge of murder and the jury could not agree on a verdict. After the jury foreman fell ill, the jury was dismissed and another trial was ordered. This time Kewish's advocates insisted he was not guilty by reason of insanity. It was, thought the jury, tantamount to an admission of guilt. After considerations lasting only an hour, the jury duly found him guilty and he was sentenced to death.

Under Manx law, Queen Victoria was required to indicate whether or not she wished the sentence to be commuted to life imprisonment, and she became reluctantly embroiled in the case. In Britain she was kept at arm's length from capital cases, only called on to dispense mercy on the recommendation of government ministers. On the face of it, Kewish had killed an older man for personal gain. There were few grounds for clemency, and the Queen was advised to give his hanging the go-ahead. So unhappy was she with her role in the affair that she insisted the law be changed so that she never again had to decide whether a subject lived or died.

After executioner William Calcraft made his way to the island, Kewish was hanged behind closed doors at Castle Rushen in Castletown. Although the Isle of Man did not abolish the death penalty until 1993, no one else since suffered the same fate there, with all death sentences being commuted to life imprisonment.

IN THE MIDDLE OF THE 1870S THE GREAT LAXEY MINE WAS ONE OF THE RICHEST IN BRITAIN

Early transport enthusiasts were intrigued by the island's travel options, which ranged from the curious to the workaday. By 1874 there was a steam train service running on a 3-foot gauge that covered some 50 miles of the island in three separate lines. Almost immediately there were extensions to it, but a decline in the mining industry left parts of the line in the doldrums. Competition from the electric railways was also an issue. However, the service continued to run in a truncated form.

Meanwhile another entirely industrial railway was at work on the island, with a diminutive 19-inch gauge. It was laid in the Great Laxey Mine, once a centre of lead and zinc exploitation. In the middle of the 1870s the Great Laxey Mine was one of the richest in Britain. It possesses the world's largest working water wheel, in grand testament to its previous significance. It wasn't until 1877 that ponies used for hauling wagons out of the mine were replaced by two small steam locomotives, endearingly named Ant and Bee, built in Poole, Dorset. An electric railway was also opened on the island in 1893, with its owners deciding the rough terrain between the Manx capital, Douglas, and Ramsey to the north would be best combated by a double track and electric power.

At the turn of the century railway companies were experiencing financial hardship that often led to company collapses and business takeovers. The Isle of Man Tramways and Electric Power Company was no exception, and it failed in 1900 while still a juvenile. Fortunately another company stepped in to save the services, which by now had gone beyond a mere tourist attraction into being a welcome and necessary part of island transport, albeit one that could boast glorious coastal and countryside views. The Manx Electric Railway remains in service with trams that run on rails and are attached to overhead wires.

The dust had long settled on railway mania when the Snaefell mountain railway was opened in 1895. A narrow-gauge railway powered by electricity, it still runs from the former mine site at Laxey to the mountain top. Its target customer was always the holidaymaker.

To combat the gradients the mountain railway used a braking system created by John Barraclough Fell. He helped to build the

The Snaefell Mountain Railway.

Furness & Whitehaven Railway before moving to Italy in the 1850s. The Fell Centre Rail System incorporated a third rail which was gripped by added drive wheels on the locomotive. There was also a special brake van to slow the progress of the train when it was heading downhill. His design was patented in 1863 and he continued experiments in England before the system was used on an Alpine railway in 1867. It was a particularly welcome development while tunnelling – the other option for mountain railways – remained a slow process.

His son, G. Noble Fell, brought the system to Snaefell where it is still visible today. It seems he was overcautious with the railway coping independently with the gradients. Nonetheless, the central rail was kept in case of emergency braking, another visible testament to great Victorian technology.

JOURNEY

5

Bradshaw's Ireland

Dublin to BELFAST

As in mainland Britain, the late nineteenth century brought railway fever to Ireland. For Victorian travellers arriving by ferry at Kingstown (now Dún Laoghaire) the chance to explore a country so rich in myth and mystery must have been truly mouthwatering. Certainly George Bradshaw thought so in his 1866 *Descriptive Railway Handbook of Great Britain and Ireland*, telling readers:

> **The entrance into the bay of Dublin unfolds one of the finest land and sea prospects ever beheld. On the right is the rugged hill of Howth, with its rocky bays, wanting only a volcano to render the surrounding scenery a fac-simile of the beautiful bay of Naples; whilst, nearer to the eye, at the extremity of a white line of masonry just fringing the sea, the light-house presents its alabaster front. On the left are the town of Dalkey with its romantic rocks, mutilated castles, Martello towers, elegant villas and the picturesque town of Dunleary, whilst behind is seen a line of parks and plantations, above which the mountains of Wicklow ascend with the greatest majesty.**

By 1866 Ireland had amassed a rail network of around 1,000 miles, driving an economic boost which 20 years previously, at the height of the Great Famine, would have been hard to envisage. Many rural towns had become accessible to tourists and trades-folk, and the east coast line between Dublin and Belfast, via Drogheda and Dundalk, opened up the great Victorian cities of Dublin and Belfast to commuters.

From Belfast it was soon possible to reach the dramatic Antrim coast and the historic port of Derry, aka Londonderry, with comparative ease. A tourist travelling from, say, Dublin to Derry could take in some of the marvels of nineteenth-century engineering, such as the Boyne and Craigmore Viaducts, while experiencing some of the unexplained, weird, wonderful and plain bizarre sights of rural Ireland.

Before following one of Bradshaw's Irish journeys, it's worth romping briefly through the country's railway history to consider how

PREVIOUS PAGE: **A view of the town of Cahirciveen on the south-west coast of Ireland.**

LEFT: **The railway line running along the coastline of Dublin Bay.**

OVERLEAF: **The Craigmore Viaduct, near Newry.**

its great designers and entrepreneurs ensured the future of steam travel, despite formidable economic, bureaucratic and geographical barriers.

The first track opened to the public in 1834 and ran for some six miles from Kingstown into the heart of Dublin. It was designed to provide an affordable mass-transportation system for commuters, and its success, just ten years after England's first railway, helped concentrate the minds of investors and financiers on the potential for new lines across Ireland.

The main instigator of the Dublin & Kingstown Railway was a leading Dublin stockbroker, James Pim, who became convinced of its viability by virtue of his own daily commute from Monkstown, above Kingstown harbour, into the city. Pim made the journey on a road teeming with horse-drawn carriages (the surface could be a mudbath in winter and a lung-busting dustbowl in summer) and reasoned that a reliable rail link would be a welcome alternative. Like many of the English railway promoters he was a Quaker, and believed strongly that a proportion of his wealth should drive progress for the public good. His connections with Quaker entrepreneurs in key ports such as Dublin, London, Bristol, Liverpool and Cardiff provided the contacts to make it happen.

The proposed railway was part funded through a £75,000 loan from the British government's Board of Works. However, the Parliamentary Bill underpinning the project ensured that financial risk was placed squarely on the shoulders of the private sponsors, the 'Gentlemen, Merchants, Traders, Freeholders and others of Dublin ... subscribers to an intended Company to be called The Dublin and Kingstown Railway Company'. Together these financiers put up £100,000 on the understanding that their personal assets would be forfeited in the event of bankruptcy.

Surveys commissioned by the directors showed there were 70,000 road journeys made per month between Kingstown and the Irish capital. Of these, 48,000 were by various types of horse-drawn carriage, 6,000 by horse, and the remainder in an eclectic selection of farm carts, gigs and 'jingles'. Averaged out per day, this translated to 2,300 journeys – a level of demand the Dublin & Kingstown Railway calculated would generate around £30,000 worth of income on operating costs of £10,000. The railway would run seven return journeys per day with regular stops en route and with enough carriages

to accommodate 300 people. Moreover, the new port of Kingstown was expected to create plenty of extra business through freight, mail and cross-channel tourism, a prospect seized upon in one early promotional notice:

> *The railway with its trains of carriages in rapid motion, will form a novel and not uninteresting foreground ... Kingstown will become a resort to which all classes will be attracted by the opportunity for the enjoyment of healthy exercise amidst a pure atmosphere and romantic surroundings.*

Of course, for the government the new line meant much more than an easy ride for commuters, a jolly day out for trippers and a profit-generator for business. The War Department understood very well that the railway could move troops quickly through Kingstown and on to Dublin in the event of an armed Irish rebellion similar to the uprising that had occurred in 1798. Its strategic importance was further heightened by fears that the French might try to invade Ireland as a stepping stone to attacking England.

In addition to Pim, the D&KR's key movers and shakers included James Perry, a financier with experience of canal construction, engineer Thomas Bergin, who was appointed general manager, and the Irish-born engineer Charles Vignoles, whose appointment was demanded by the Board of Trade to ensure taxpayers' cash was spent wisely. But above all the company needed a hands-on, experienced contractor who could drive the project forward and prove that steam was the future. In William Dargan they found the very man. Not for nothing would he be hailed as Ireland's 'King of the Railways'.

Born in 1799, Dargan was the son of a tenant farmer from rural Carlow. His natural aptitude for mathematics, a sound English education and good family contacts helped him begin a career as a back-office junior with a Dublin surveying company. Later he travelled back across the Irish Sea to work on George Stephenson's plans for the *Rocket* and was soon hired to help on Thomas Telford's London-to-Holyhead rail link. In 1820 he started surveying for another Telford project – the road to Holyhead, Anglesey. After this Dargan returned to Ireland, set up on his own account and completed various small engineering projects before making his name through the construction

PREVIOUS PAGE: ***Granite Pavilions and Tunnel Entrance at Lord Cloncurry's Demesne of Maratimo near Blackrock, Kingstown Harbour in the Distance*, engraved by S. G. Hughes, 1834.**

RIGHT: **Travelling on the Downpatrick Heritage Steam Railway.**

ABOVE: **William Dargan (1799–1867).**

of the Dublin-to-Howth road, an important arterial route north of the capital, and the hugely impressive Ulster Canal link between Belfast and Lough Erne. But railways had become his passion, and he seized the chance to submit a tender for the D&KR project.

By now Dargan had carved out an enviable reputation as an entrepreneur of vision. He became popular with his workforce because he paid by results (a scheme that was much more lucrative for the labourers than the miserable day-rates offered by some construction companies) and made sure wages arrived punctually. His navvies told other workers of his reputation for fairness, with the result that Dargan had access to a reliable, able and willing labour force, making him well placed to meet the huge demand for track expansion which lay ahead.

IT'S HARD TO OVERSTATE THE COLOSSAL TASK, IN MAN HOURS, WHICH FACED THE EARLY RAILWAY LABOURERS

It's hard to overstate the colossal task, in man hours, which faced the early railway labourers who weren't working under Dargan's enlightened management. Initially hours were long and pay poor, a state of affairs which led to simmering resentment and ultimately, in June 1833, a riot. Some of the 2,000 navvies employed on the D&KR clashed with police and the entire site ground to a halt until Dargan persuaded Pim and Bergin to introduce his piecework system of remuneration. Typically, this meant paying a man by the number of spoil-filled wheelbarrows he shifted.

Industrial action was by no means the only thorny issue. Negotiating land purchases with wealthy estate owners was a management horror story, given that the incumbents held the cards. In later years George Bradshaw's readers bound for Dublin would admire the superb Italianate Bridge above the line at Blackrock and, nearby, an elegant folly on the shore which appeared to be a bathing hut modelled on an ancient Greek temple. Both were built by the D&KR at the behest of landowners – but hardly willingly.

The problem for the railway was that hearings before an Irish Land Jury to decide fair compensation were slow and cripplingly expensive. Achieving a swift out-of-court settlement was therefore preferable, but this meant kowtowing to the landowners and to two in particular – Baron Cloncurry and the Reverend Lees. Cloncurry, an eccentric Irish nationalist who had served time for his anti-government activities, squeezed £3,000 out of the D&KR, plus an additional bridge across the line, built 'in the best Italianate style', so that he could walk his dogs down to a private bathing area. He also required

A coloured drawing of the steam locomotive *Hibernia*, 1834.

a granite bathing hut modelled on a Greek 'pleasure-dome for men'. Lees's demands were limited to cash but he also forced a hard bargain, winning an exorbitant £7,500 in compensation.

Construction work finally started in 1833 and the service was launched, half an hour late, at 9.30 a.m. on 17 December 1834, when the Manchester-built locomotive *Hibernia*, pulling packed carriages, steamed out of Dublin for the 19½-minute ride to Kingstown Harbour. Historically, this truly was a great train journey, as the *Dublin Evening Post* reported the following day:

> *This splendid work was yesterday opened to the public for the regular transmission of passengers to and from Kingstown, and the intermediate stage at the Black-rock. Notwithstanding the early hour at which the first train started, half-past nine o'clock, the carriages were filled by a very fashionable concourse of persons, and the greatest eagerness was manifested to witness the first operations of the work. Up to a quarter past five the line of road from Merrion to Salthill was thronged with spectators, who loudly cheered each train that passed them.*

A view of Killiney Bay, Co. Dublin, looking towards Bray Head.

The reporter is mildly critical of the railway's method of boarding passengers – 'much confusion was occasioned at starting up by the want of proper arrangement' – but praised precautions taken to prevent accidents. These included a team of line marshals and mandatory slow starts for the engines. 'Although there could not have been less than from three to four thousand persons upon the railway during the day, we are happy to state that these very necessary precautions were attended with the desired effect.'

However, according to *The Times* of Monday 22 December, those travelling on the train were at substantially greater risk than those watching from trackside:

> *All the machinery works well as yet, except in one particular: the springs are not sufficiently elastic to prevent sudden shocks when the carriages stop. Three or four gentlemen had one occasion to-day* [sic] *their heads knocked against each other and the carriage doors and severe contusions were the consequence. A county of Kildare gentleman's head was laid open. The majority had, however, hard Irish heads, and did not mind a few knocks.*

Health and safety laws were rather less rigorous in those days, and *The Times* correspondent cheerily concludes: 'The weather is delightful for December, and a few broken heads does not throw much damp on a scene of Irish amusement where everything else goes well.'

The success of the D&KR led to demands for a line extension south-east, and in 1844 Dargan obliged with the Dalkey Railway. Among the guests at this opening ceremony was Isambard Kingdom Brunel, arguably the greatest of all Victorian engineers, who took the opportunity to reveal an ambitious new venture to the D&KR Board. Brunel explained that he was planning a new line across South Wales, servicing a sea ferry between Fishguard in Pembrokeshire, South Wales and Rosslare, the closest port to Wexford in Ireland's far south-east. Logically, he needed a rail link from Wexford to Dublin. Would the Board consider a joint venture to achieve this? The proposal was duly agreed and the Waterford, Wexford, Wicklow & Dublin Railway was born. Work on the route began at various locations in 1847 but construction progressed agonisingly slowly due to geological problems around Bray Head. It wasn't until 1854 that the stretch from Dalkey to Bray was finished, while the Wicklow section didn't open until the following year.

The boost to Bray's economy was a game-changer for locals. Suddenly it was possible for people to live in the town yet commute daily to Dublin; within 20 years of the line's opening in 1854 the area's population rose from 4,151 to 6,504. Railway trippers would pour down from the city or Kingstown ferry, perhaps conscious of Bradshaw's exhortations:

> **The stranger would do well to make this place his headquarters for a few days, being most beautifully situated, and in the very heart of scenery the most attractive. In the town are the remains of a castle (now used as a barrack) and a race course. There is also a pretty lake and a river abounding with trout. Close at hand is Kilruddy *[sic]*, near the Sugar Loaf, seat of the Earl of Meath. Bray Head is within twenty minutes walk of the station and is well worth a visit. Its excellent winding path, cut in the cliff, affords the tourist a magnificent view, embracing the Hill of Howth, Dalkey Island and Killiney Bay, with the railroad below, and the ocean washing the base of the 'head'.**

The prosperity being bestowed on previously isolated towns was not lost on the rest of Ireland (although this sentiment faded as people became more cynical about the railway investment 'bubble'). In 1852 half the electorate of Cashel, Co. Tipperary, pledged to support *any* prospective Member of Parliament – irrespective of his politics – who could provide a rail link to the town. It seems the public took the view that establishing a railway was not that hard. Back in Bray, Dargan and Brunel would have begged to differ.

Scenic it may have been (and still is) but the geology and geography governing their next section of track, south past Bray Head, was the stuff of nightmares for an engineer. There are competing theories as to why they persisted with a coastal line rather than a simpler, inland route to Wicklow. Some say it was to tempt tourists with the glorious

IN 1852 HALF THE ELECTORATE OF CASHEL ... PLEDGED TO SUPPORT *ANY* PROSPECTIVE MEMBER OF PARLIAMENT ... WHO COULD PROVIDE A RAIL LINK TO THE TOWN

RIGHT: **A geological map of Ireland, drawn and engraved by John Emslie, c. 1855.**

OVERLEAF: **The accident at Bray Head in 1867.**

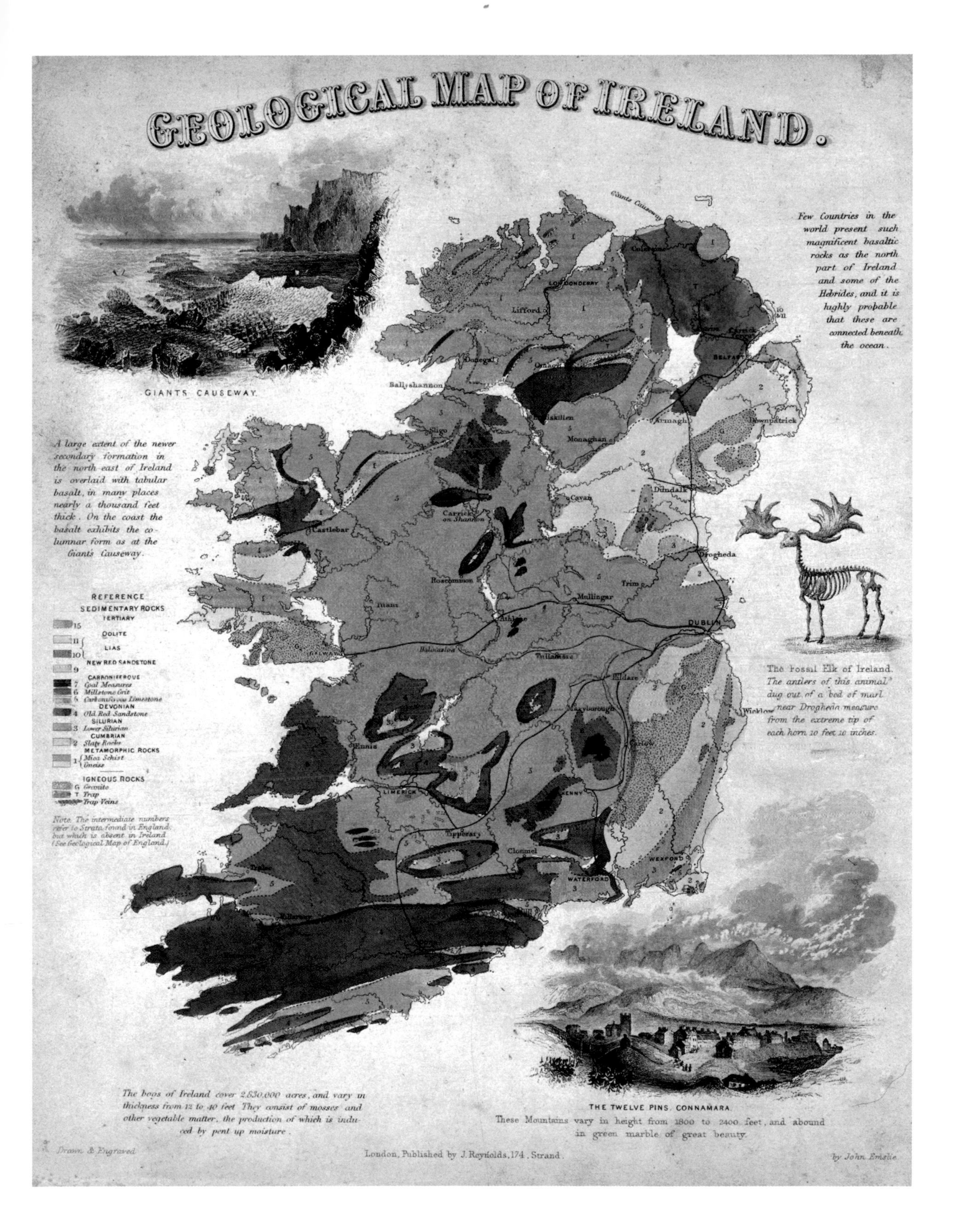
GEOLOGICAL MAP OF IRELAND.
GIANTS CAUSEWAY.
Few Countries in the world present such magnificent basaltic rocks as the north part of Ireland and some of the Hebrides, and it is highly probable that these are connected beneath the ocean.
A large extent of the newer secondary formation in the north east of Ireland is overlaid with tabular basalt, in many places nearly a thousand feet thick. On the coast the basalt exhibits the columnar form as at the Giants Causeway.
REFERENCE
SEDIMENTARY ROCKS
TERTIARY
OOLITE
LIAS
NEW RED SANDSTONE
CARBONIFEROUS
DEVONIAN
SILURIAN
CUMBRIAN
METAMORPHIC ROCKS
IGNEOUS ROCKS
The Fossil Elk of Ireland. The antlers of this animal dug out of a bed of marl near Drogheda measure from the extreme tip of each horn 10 feet 10 inches.
The bogs of Ireland cover 2,830,000 acres, and vary in thickness from 12 to 40 feet They consist of mosses and other vegetable matter, the production of which is induced by pent up moisture.
THE TWELVE PINS, CONNAMARA.
These Mountains vary in height from 1800 to 2400 feet, and abound in green marble of great beauty.
Drawn & Engraved
London, Published by J. Reynolds, 174, Strand.
by John Emslie.
LONDONDERRY
BELFAST
DUBLIN
LIMERICK
WATERFORD
WEXFORD
Drogheda
Mullingar
Castlebar
Sligo
Ennis
Clonmel

cliff views; others claim Lord Meath donated land around Bray Head to ensure the vista from his Kilruddery estate wasn't spoiled. No doubt there was also a part of Brunel and Dargan which simply relished the challenge; critics of the new line often claimed that 'Bray Head would never be conquered' by the railway.

As it turned out the naysayers would be proved wrong, though at times it was a close-run thing. The rock on these cliffs is Cambrian and the vertical strata can easily collapse when disturbed. This made tunnelling tricky while the erosive effects of the waves and wind on the foundations produced regular landslips. Brunel's experience of constructing the Thames tunnel and Bristol's Clifton Suspension Bridge proved invaluable as work slowly progressed down three separate tunnels. But long after the line was finished four new tunnels had to be bored – the last in 1917. Each saw the track pushed further and further back from the sea (today's travellers can still see the old entrances) and a profit-sapping £40,000 was spent on sea defences inside ten years.

At one point a rock fall forced the line to be moved 10 feet. The bridging of a ravine with a 300-ft-long, 75-ft-high wooden viaduct was severely delayed when the first structure was destroyed by heavy seas. And such was the unstable nature of the cliffs that trains traversing a ledge 70 feet above the waves needed protection from rock falls; roofs were bolted into the cliff face above the track.

Against this background accidents were almost inevitable, and on 23 April 1865 a first-class carriage of the Dublin train jumped the rails in the middle of a viaduct. The engine halted but then, remarkably, pulled the carriage to safety. Less fortunate were two passengers who, two years later, lost their lives as an engine, tender and three carriages crashed through the side of a bridge and plunged 30 feet into a ravine.

Yet the Victorian can-do approach to engineering eventually triumphed. The line to Wicklow opened in 1855 and reached the copper mines at Avoca via a majestic five-arch viaduct in July 1863. From here the track was quickly extended towards Arklow and Enniscorthy, finally reaching Wexford in 1872. Brunel and Dargan had fulfilled their dream of opening the south to the rail revolution – albeit with a legacy of high maintenance.

Of course Dargan did not confine himself to contracts south of Dublin. By 1853 he had already constructed around 600 miles of track, including sections of what would become the Great Southern & Western Railway, the Midland Great Western Railway and the Great Northern Railway. The latter was itself the result of mergers between three founding companies: the Dublin & Drogheda Railway, the Ulster Railway and the Dublin & Belfast Junction Railway.

The success of the Dublin & Kingstown Railway's partnership with Brunel in the south-east could have been a useful business model for these emerging operators. The reality was very different. Each had their own plans, their own routes, their own trains and, especially, their own track gauges. There was no Regulator galvanising the various Boards into considering the necessity for national services in which trains from one company could run on the tracks of another.

This problem might never have arisen if Thomas Drummond, Under-Secretary of State for Ireland from 1835 to 1840 and a mainspring of the Irish Railway Commission, had had his way. In 1837 Drummond headed the Commission and tried to argue for a greater degree of state control in the industry's development. But his ideas were anathema to the private enterprise lobby and the result was industrial brinkmanship on a grand scale.

In Belfast many factory owners had made their fortunes from the Industrial Revolution and knew how steam technology had transformed their linen mills. They began pushing for their own rail link, and in 1836 an Act of Parliament approved the founding of the Ulster Railway Company. Its aim was to establish a line south from Belfast, via Lisburn, Lurgan and Portadown, to Armagh, thus providing the population of the Lagan Valley (seen as an untapped labour source) speedy access to the textile mills.

Construction began in 1837 with George Stephenson as Ulster Railway's consulting engineer. In fact Stephenson spent only a few days in Belfast and it was William Dargan who took on the bulk of the work. He monitored the work in his personal coach – fitted out as a mobile operations HQ – and again charmed the 3,000-strong workforce with piecework bonuses linked to the speed of advance. Within two years the line was open between Belfast and Lisburn, with return fares fixed at a shilling for first class and sixpence for second.

While the Ulster Railway pressed on to Armagh, the Dublin financiers behind the successful Dublin & Kingstown Railway were busy preparing their own plans for a direct link between Ireland's two great cities. They formed the Dublin & Drogheda Railway Company and unveiled a route which headed straight up the east coast. The Dubliners saw no reason to divert inland to Armagh – it was not in their commercial interests to deliver passengers to the Ulster Railway – but their Belfast rivals treated their proposal with disdain, claiming it would reduce overall demand, and therefore the profitability of both lines. The linen merchants also genuinely feared that the hills between Dundalk and Newry would prove too costly an engineering challenge.

Deadlock followed. The 1838 Railway Commission preferred the inland route via Armagh, but wrangling continued, leading to a two-year delay for the D&DR. The latter eventually appointed William Dargan as its main contractor with Professor John MacNeill, first

PREVIOUS PAGE: A locomotive on the Downpatrick Heritage Railway.

BELOW: A locomotive from the Irish Railway collection at the Ulster Folk and Transport Museum.

professor of engineering at Trinity College Dublin, as consultant. And so was seeded the genesis of a new problem.

MacNeill advised a gauge of 5 ft 2 in, arguing that this would make the track cheaper to build. However, the 1838 Commission had already recommended 6 ft 2 in and the Ulster Railway had dutifully followed that guideline. To further complicate matters, the D&KR had mirrored George Stephenson's *Rocket* with a 4 ft 8½ in gauge. It seemed Ireland was destined for the same compatibility problem as had bedevilled the early rail industry across the water. It was left to the Board of Trade to sort things out.

The Board appointed a well-respected arbiter, Major-General Sir Charles Pasley, the Inspector-General of Railways and a former head of the School of Military Engineering, Woolwich. Pasley quickly ruled out Brunel's suggestion of a 7 ft gauge and then considered the Stephensons' advice that a compromise should be struck somewhere between 5 ft and 5 ft 6 in. Pasley presumably saw little reason to prolong the debate. He chose 5 ft 3 in, one of those rare imperial lengths which can be converted precisely to a metric round figure: 1,600 mm.

This caused few problems for the Dubliners as the Dublin & Drogheda Railway was still at an early stage of construction. But the Ulster Railway faced a £20,000 bill to replace its original gauge, and in 1843 successfully argued to the Board of Trade that other companies should pay it compensation. A clause covering this appeared in several Railway Acts from 1850 onwards. To minimise disruption the Ulster Railway built a 5 ft 3 in line alongside its redundant 6 ft 2 in original, and completed a full switchover in September 1847. Incidentally, the Dublin & Kingstown wasn't changed until 1857, no doubt because at £38,000 the cost was not easily recouped. Interestingly the 'Irish Gauge' is now only found in three Australian states – Victoria, New South Wales and South Australia – and Brazil.

In 1844 the D&DR completed its track to Drogheda. Four years later, the UR reached Armagh. Around 56 miles of glorious, green Irish countryside lay between the two, with the result that a new player entered the game – the Dublin & Belfast Junction Railway. This was incorporated by an 1845 Act of Parliament with the declared intention of extending the D&DR line from Drogheda to the key UR station at Portadown, north-east of Armagh. And once more William Dargan got the lion's share of the contract.

THE 'IRISH GAUGE' IS NOW ONLY FOUND IN THREE AUSTRALIAN STATES – VICTORIA, NEW SOUTH WALES AND SOUTH AUSTRALIA – AND BRAZIL

Work started at numerous locations along the proposed route but Dargan took personal charge of the section south of Portadown across the difficult marshland area towards Mullaghglass and Goraghwood. The Newry Valley was his next major challenge; the quarter-mile-long Craigmore Viaduct was needed to span the River Camlough and connect to the important coastal town of Dundalk. But a still more complex engineering project lay ahead – the bridging of the River Boyne further south at Drogheda. Dargan used designs by John MacNeill for both the Craigmore and Boyne Viaducts, though not without some controversy. When they were completed in 1849 and 1855 respectively these bridges ranked among the wonders of the railway world and were landmark attractions for Victorian tourists.

With the Dublin to Belfast link finally complete there was a new impetus for railway entrepreneurs to build connecting branch lines. From Belfast this allowed Bradshaw's late Victorian travellers to embark on a series of rambling detours, and our route takes in a couple of these – the Belfast & County Down Railway south-east to Downpatrick, supposedly the site of St Patrick's grave, and the Carrickfergus & Larne Railway north to Whitehead and the extraordinary Gobbins Path. But the main route continues on north-west, along the old Belfast & Northern Counties line, via Ballymena and Coleraine to Londonderry.

With the end of the Great Famine in 1849, and with track construction proceeding apace, William Dargan decided it was time to relaunch Ireland as a new industrial powerhouse. Inspired by London's Great Exhibition of 1851, he kick-started plans for the 1853 Dublin Exhibition by depositing £30,000 with various organising committees, and later by advancing a further £70,000 in loans and guarantees. Dargan lost around £20,000 on the venture, but achieved his aim. More than 100,000 people attended the exhibition, including Queen Victoria, Prince Albert and the Prince of Wales, and there were claims that 1 million visitors were attracted to Dublin that year. On 29 August 1853 the Queen visited Dargan and his wife, Jane, at their Mount Annville house where, according to the *Illustrated London News* of 10 September, she offered Dargan a baronetcy in recognition of his contribution to the railways. He politely declined.

Dargan's final years were devoted to extending Dublin's rail link south to Wexford. He also dabbled in the tourism business, building

OVERLEAF: Sackville Street and O'Connell Bridge, Dublin.

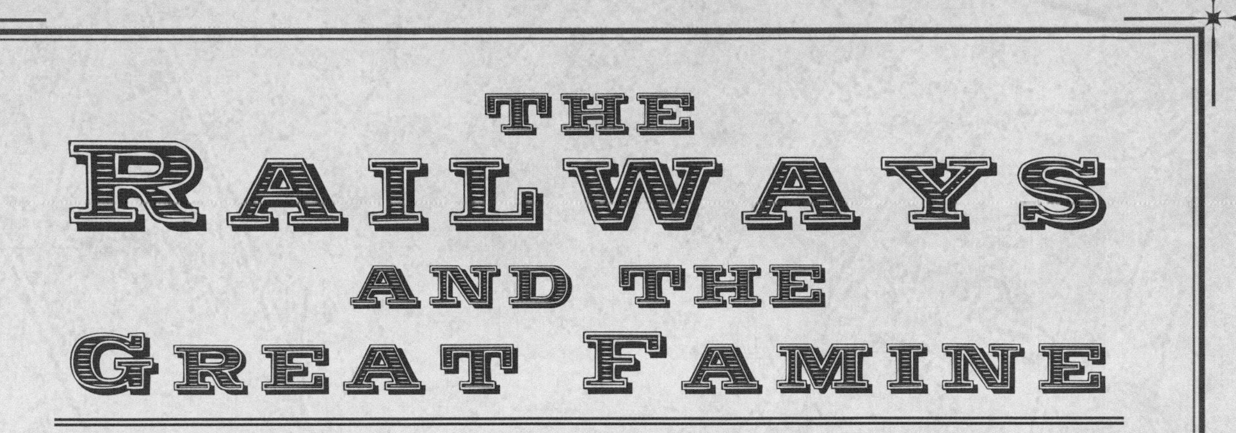

The Railways and the Great Famine

The fact that Ireland built such a vast railway network at a time when the country was suffering appalling deprivations from the Great Famine of 1845–9 is a tribute to the indomitable Irish spirit. But from another perspective it also heightens the atrocious failure of government, in both Westminster and Dublin Castle, to grasp the disaster of genocidal proportions that was unfolding. This is not a book about the Famine but it is worth highlighting what the shiny new Irish railways did, and did not do, for their starving people.

Historians differ over the exact number who died – statisticians put the figure anywhere between 775,000 and 1.5 million – but whatever the numbers the human tragedy is beyond the imagination of most Westerners today. Thousands queued on the streets at soup kitchens, the sick and painfully thin besieged rural hospitals – 'skeletal armies', as the Irish historian Robert Foster described them – while in the far west of the country entire families walled themselves into cabins and cottages to await death. Irrespective of who was to blame, the Great Famine remains perhaps the most shameful peacetime event in recent British history.

The irony, of course, is that the railways offered the greatest potential for mass transport of food which Ireland had ever seen. Instead, they became the conduit for a mass exodus. Some 1.5 million people are thought to have emigrated during the Famine years, and although the government did provide some food relief, political leaders saw the railways more as a means of providing extra employment, allowing labourers to feed their families, than a food distribution network. Certainly the great campaigner for Catholic emancipation and Lord Mayor of Dublin, Daniel O'Connell, pressurised the companies into hiring extra labour, and there is evidence that Dargan took on more men than he strictly needed to in the worst-affected areas.

Irrespective of blame the Great Famine remains perhaps the most shameful peacetime event in recent British history

and running the Royal Marine Hotel at Kingstown. Presumably this magnificent landmark building, which opened in 1865, didn't quite hit Bradshaw's deadline for the 1866 *Descriptive Guide* – he mentions only Rathbone's Hotel as a berth for weary ferry travellers – but it soon developed a reputation for luxury and extravagant cuisine. When Queen Victoria visited Ireland on 4 April 1900 she apparently headed straight off the boat and into the Royal Marine Hotel for a 16-course breakfast.

Sadly, Dargan was seriously injured falling from a horse in 1866. He never fully recovered, his financial affairs unravelled and he ended direct involvement in his railway projects. He died on 7 February 1867 and was buried in Dublin's Glasnevin cemetery. In 2004 a new cable bridge for the Dublin Light Railway was named Dargan Bridge in his honour.

There seems little doubt that the coming of the railways made a huge contribution to Dublin's prosperity. Bradshaw notes:

> **The appearance of Dublin is very much improved of late years. Streets have been widened, new squares skilfully laid out, and many public monuments freed from buildings which concealed their beauties.**

He is captivated by the view from Carlisle Bridge, the ship-cluttered Liffey, the Bank – 'the most perfect building in Dublin'– and a dozen or more public edifices which would no doubt interest the architecturally literate rail traveller. However, the one attraction which really did inspire the masses to take to the rails gets only a passing reference, alongside the Phoenix Park Barracks: Dublin Zoo.

The Victorian public loved zoos, and railway managers realised that here was an excellent opportunity to fill trains on high days and holidays. Walsh's train timetable was advertising Dublin Zoo in 1848 (although with a 6d standard entry fee it was still an expensive day out) and Penny Sundays were introduced to attract working-class families. Still, it wasn't until the Dublin Exhibition that zoo visitor numbers really took off, and 120,000 passed through the gates that year. In 1855 the zoo acquired its first pair of lions; they bred for the first time in 1857 and soon the lions of Dublin, particularly a superb specimen

Tylers' BOOTS ARE THE BEST
HATCH ST

named Charlie, became a world-famous attraction – so much so, it was claimed that the lions pushed up railway share prices.

It says as much in a colourful newspaper report on the visit to Dublin in 1878 by General Ulysses S. Grant, the renowned commander of the Union forces during the American Civil War and former President. It was filed with the minutes of the Royal Zoological Society of Ireland in December that year.

BALBRIGGAN WINKLES WERE MUCH IN DEMAND AND SOLD IN BULK AS FAR AWAY AS LONDON'S BILLINGSGATE MARKET

> *General Grant (former US President) visited the Zoo and was observed to smile. This remarkable event occurred during his visit to the Zoo … Secretary [the Rev. Professor Samuel Haughton] exhibited his tigers, jaguar, pumas, leopards and even the new lioness … the general neither spoke nor smiled at any of their performances and at length stood impassive, lighting a fresh cigar opposite the cage of the celebrated lion 'Charlie' the gem of the collection.*
>
> *Hereupon the secretary, changing his manner of deferential courtesy, which he had hitherto maintained, into one of more familiarity, said: 'General, that lion's name is Charlie, he is three inches higher in the shoulder than his grandfather, who came from South Africa. I have reared ninety-six of that old lion's children and grandchildren in these gardens and I reckon Charlie is the biggest of them. I am now going to tell you something which you will not, perhaps, believe: that lion was the cause of the shares of the South Western Railway rising 2 per cent in the open market. When it became known in the United States how large that lion had grown, the Transatlantic passengers all commenced landing at Queenstown [Co. Cork] instead of proceeding to Liverpool by sea. They now travel by the GSW line from Queenstown to Dublin and on arrival drive out to the gardens to see the big lion. Comfortable seats are provided for them in the lion house where they sit for hours smoking silently and admiring the proportions of the vast brute.'*

With Bray, Kingstown and Dublin disappearing into the distance, the route heads due north, along the old Dublin & Drogheda Railway via the fishing port of Balbriggan. Bradshaw has little to say of Balbriggan, presumably calculating that readers will find limited interest beyond the 35-ft-tall lighthouse and 600-ft-long pier. There's a brief mention for the

The old lighthouse at the entrance to the harbour at Balbriggan, Co. Dublin.

local industries – stocking, linen, tanning, muslin and embroidering trades – but nothing of the railway's important links with the local fishing industry.

It was a truly symbiotic relationship. The fleet's catch provided regular freight business and during the herring season between October and January so many boats based themselves at Balbriggan that it was said you could walk across the harbour without getting wet. The local kipper factory flourished, as did the trade in winkles – a favourite Victorian delicacy. Winkles were gathered in sacks at the harbour and then hauled up the hill to be weighed at the station. Once the sacks contained a couple of tons of shellfish they would be loaded onto the next train and transported to market. Balbriggan winkles were much in demand and sold in bulk as far away as London's Billingsgate market.

The railway played one other, unintentional role in supporting fishermen. As trains crossed the viaduct alongside the harbour, drivers were instructed to give a loud blast on the whistle. This helped guide boats back to harbour in fog or at night, and was a practice which continued right up to the 1950s, when new navigational equipment became available.

From Balbriggan Bradshaw's traveller passed through what the guide describes as an 'undulating, rich and highly cultivated plain'. While the countryside is agreeable enough, Bradshaw is scathing about the living standards of the poor, warning readers:

> **The farms are often very extensive but the farm houses, except when they belong to large proprietors, are in general wretched huts; and the houses of the humbler classes are nothing more than mud hovels.**

Thankfully his mood brightens when he crosses the River Boyne, perhaps because it offers an opportunity to enlighten readers on one of modern Ireland's bloodiest and most significant events, the 1690 Battle of the Boyne, which effectively ensured Protestant dominance of Irish public life for generations to come. From Bradshaw's point of view this location demands the retelling of Irish history's most withering put-down.

The battle was fought between the deposed Catholic monarch King James II (James VII of Scotland) and his son-in-law and successor, the Protestant ruler of the Netherlands, William of Orange (William III). James was the last Catholic to rule the combined British kingdoms but was distrusted by Protestant leaders as being too pro-Pope and (possibly even worse) pro-French. They invited William to claim the throne, and in 1688 his invasion forced James to flee to France.

The following year James landed in Ireland with the intention of reclaiming his birthright. But his inexperience as a commander shackled his mainly Irish Catholic army, who, though mostly raw recruits, were by no means destined for defeat. Many of his soldiers were furious at James's lack of personal leadership (he stood on a hill out of the way, in contrast to William, who led his troops from the front) and Irish commanders were aghast when the order came to retreat from the banks of the Boyne. James, with a small escort, was first to head for safety – earning the Gaelic nickname *Seamus an Chaca* (James the

British cavalry charge through the waters of the Boyne at the Battle of the Boyne, 1690.

Shit) – and his French cavalry proved similarly keen to disengage. While many Irishmen regrouped to fight on for another year, James rapidly headed south, first to Dublin, then to Duncannon harbour, bound for exile in France.

In his description of the battle and aftermath, Bradshaw writes:

> 'Change leaders', said the beaten Irish, 'and we will fight the battle over again' but their despicable sovereign made off as fast as he could to Dublin where he met the Duchess of Tyrconnel. 'Your countrymen run well, madam,' he said. 'Not quite so well as Your Majesty,' said the lady, 'for I see you have won the race.'

The Dublin and Belfast Viaduct across the Boyne near Drogheda.

For the Victorian traveller to Drogheda there was obviously little tangible evidence of the battlefield, although Bradshaw recommends a carriage drive through King William's Glen to get a flavour. But in terms of hard-core sightseeing the Boyne Viaduct must have more than made up for this. Quite apart from the technical achievement, the bridge was both a masterpiece of design and the final link in the Dublin–Belfast rail link. It said to the world that Ireland was in the fast lane of industrial development.

Work didn't begin on the viaduct until 1850, delayed by the continuing shadow of the Famine and a protracted dispute over the design calculations. The 98-ft-high bridge had 15 stone arches bearing 1,760 feet of track; its 226-ft-long central iron truss stood on piers sunk 30 feet into a morass of alluvial clay, tons of which had to be removed by chain bucket. With such a massive commitment in terms of labour and materials it was essential that planning drawings were error-free.

Professor John MacNeill (now consultant engineer to the Dublin & Belfast Junction Railway) had handed in his specifications but the company's chief engineer, James Barton, questioned the load-bearing calculations and it was left to the Department of Engineering at Trinity College to adjudicate. No doubt choosing words carefully, the Department backed Barton and recommended a strengthening of the latticework, which was then the longest stretch in the world. It was redesigned so that two locomotives and carriages, with a combined weight of 1,000 tons, could pass simultaneously on the central point of the viaduct. While such an occurrence was deemed unlikely in the extreme, Board of Trade railway inspectors were known to have a fearsome reputation for strict safety margins. Ultimately this was not misplaced; the Armagh Rail Disaster (see below) was a classic example of the tensions between company profits and the interests of passengers.

THE BOYNE VIADUCT WAS BOTH A MASTERPIECE OF DESIGN AND THE FINAL LINK IN THE DUBLIN–BELFAST RAIL LINK

When the Boyne Viaduct opened for business many rail travellers were entirely unconvinced by its stability, even though tests proved the structure was accurate to one third of an inch – an extraordinary achievement. It was a view no doubt encouraged by carriage drivers who had made a good living ferrying Dublin and Belfast-bound passengers across the river via a road bridge. Public concern was heightened at the 1855 opening by the sight of wooden scaffolding (used for construction rather than load-bearing) still cloaking the sides of the new bridge.

These fears soon passed and the viaduct was hailed by many as a wonder of the industrial age. Yet its completion was never a cause for wild celebration. The famine years were not yet over and the railway was seen by some critics as an instrument of an uncaring government. There were even reports of starving farmworkers in the far south-west attacking railway employees.

Understandably, Bradshaw makes no mention of this and from Drogheda escorts his readers north through the 'inexpressibly beautiful' countryside of the Boyne, the rural stations of Beauparc and Navan and on to Kells, Co. Louth, an important centre of Irish Bronze Age architecture and culture.

One of the smallest counties in Ireland, it abounds in those rude vestiges of antiquity which consist of earth-works, chiefly designed for sepulchral purposes, or acting as places of defensive habitation. Cromlechs, and other relics of anti-Christian ages,

although much lessened in number within the last century, are still numerous, and in some instances extremely curious.

Among the most curious is the Proleek Dolmen, a short carriage drive from the railway, which stands in the grounds of the Ballymascanlon House Hotel (built in the 1860s). The two upright or 'portal' stones mark the doorway to a 4,000-year-old burial chamber. They support a 40-ton capstone known as 'the Giant's Load' on the basis that only a giant could have placed it there (the giant concerned was apparently Parrah Boug MacShagean, from Scotland). This was all good stuff for a Victorian public fascinated by mystical matters. According to local folklore, anyone throwing onto the capstone a stone which stayed in place would be married within the year. Giant or no giant, it had to be good for the tourist trade.

From Louth, the journey north continues through the seaport of Dundalk and on to Newry, where the guide notes that a yew tree in the grounds of a thirteenth-century monastery was planted by Ireland's foremost patron saint, St Patrick. By Bradshaw's time Newry had become an important railway hub, providing both a service to Belfast via Portadown and a link between Armagh to the north-west and the seaside town of Warrenpoint to the south-east. It was along this latter stretch of track, 2½ miles out from Armagh, that a tragic combination of circumstances led to the deaths of 80 people and changed the operation of railways across Europe forever.

On 12 June 1889 a special excursion train, booked by Armagh's Methodist Church Sunday School, set out on the 24-mile day trip to Warrenpoint. Organisers had planned for 800 people and the engine house at Dundalk calculated that 13 carriages would be required, pulled by a four-coupled locomotive (i.e. with two pairs of driving wheels). In the event they sent two extra carriages with instructions to driver Thomas McGrath that these should not be used. Neither McGrath nor his Dundalk managers were familiar with the steep, steady climb for the first few miles of track south-east of Armagh. The locomotive concerned, engine 86, was probably underpowered – even for 13 coaches.

Saint Patrick, the patron saint of Ireland.

Unfortunately, many more than 800 passengers turned up to join the Warrenpoint Special that day, and the Armagh stationmaster, John Foster, began issuing tickets for all 15 carriages. Evidence at the subsequent accident inquiry was confusing and contradictory; McGrath claimed he'd opposed the increase while Foster insisted the argument was over his own request for still more coaches. According to McGrath the stationmaster told him ten minutes before departure: 'Any driver that comes here does not grumble about taking an excursion train with him.' McGrath claimed that he had replied: 'Why did you not send proper word to Dundalk, and I should have a proper six-wheeled coupled engine.'

The railway's chief clerk, a Mr Elliott, who was accompanying the excursion, suggested a solution: the scheduled train following 15 minutes behind could assist the Special up the climb. Either that, or some carriages could be left behind for the later service. However, McGrath was reluctant to do this, and at 10.20 a.m. he set off with 940 people crammed into 15 carriages.

Initially it seemed the decision might be vindicated. There had been no rain to turn the rails slippery, and at full throttle the locomotive made good progress. But then at Derry's Crossing, some 700 yards from the Dobbin's Bridge summit, the train began losing momentum. It halted just 200 yards short of the top.

To stop the train rolling back McGrath applied what were known as 'continuous brakes'. These relied on a vacuum produced by the engine; if power was lost the brakes were released. It was not a system recommended by the Board of Trade, which preferred a fail-safe arrangement in which a vacuum held the brakes *off* while the engine was running. These so-called 'automatic continuous brakes' slammed on immediately in the event of engine failure.

TO STOP THE TRAIN ROLLING BACK MCGRATH APPLIED WHAT WERE KNOWN AS 'CONTINUOUS BRAKES'

With the Special stationary, guards in the two brake vans at the front and rear of the train applied their handbrakes. Some stones were used to wedge or 'scotch' carriage wheels (ineffectively, as it turned out) and the chief clerk ordered the crew to divide the train, taking on only the front five carriages to the next station at Hamilton Bawn. The engine could then return to collect the remaining ten carriages.

From the moment the fifth and sixth carriages were slackened and decoupled the continuous brakes came off the entire rear section of the train. As Thomas McGrath attempted to move away he rolled back slightly, striking the sixth carriage. This was enough to crush the makeshift stone wedges and produce sufficient momentum for the ten rear carriages to overcome the sole remaining handbrake. Inexorably the Special began rolling back – a runaway train heading straight for the 10.35 scheduled service steaming out of Armagh. It was the railwayman's worst nightmare. McGrath and his crew reversed the front section of the train and tried pursuing the carriages in an attempt to re-couple them, an act of desperation doomed from the outset.

Paddy Murphy, the driver of the 10.35 service, spotted the carriages heading towards him when they were still 500 yards away. He managed to cut his own speed to 5 mph but by now the freewheeling carriages were travelling at 40 mph. It must have been a dreadful sight, with passengers flinging themselves from the running boards and children thrown from the windows. In the inevitable collision carriages 13, 14 and 15 were obliterated, 80 people died (including 20 under the age of 15) and at least 170 were injured. Murphy survived but he never drove a train again.

Wreckage litters the hillside after the Armagh Rail Disaster on 12 June 1889.

The subsequent inquiry cast a very public spotlight on the murkier corners of the railway industry. The inspector's findings were many and varied but the upshot was a realisation that passenger safety was not a top priority, that track and signalling procedures were unacceptable, and that many railwaymen worked way above their allotted hours. In Ireland it was discovered that only one engine and six carriages were equipped with automatic continuous brakes.

Within two months of the Armagh Disaster – and despite intensive lobbying from rail companies pleading poverty – Parliament enacted the Regulation of Railways Act 1889. This made automatic continuous brakes compulsory on all passenger railways, introduced a block system of signalling (which prevented a train from entering a track area until it was confirmed clear) and required the interlocking of points and signals (to prevent false signals). Out of the tragedy emerged a determination to learn – a heartening example of Parliament placing people before profit. Hasty it may have been, but the 1889 Act ushered in the modern era of rail safety.

In Bradshaw's day Belfast was in the midst of an unprecedented era of prosperity driven by its high-tech textile industry, a fast-expanding port and the shipyards. While Dublin was the seat of government, it was Belfast which set the pace of Ireland's economy. Nowhere is this clearer than in the census figures; in 1841 Belfast's population was 70,447, compared with Dublin's 232,726. Yet within half a century Belfast was the larger city, the 1891 census recording a population of 276,114 compared with 269,716 for its southern rival.

Bradshaw gives a nod to this expansion, noting that customs duties at the port rose from £3,700 in 1805 to more than £360,000 in 1846.

> **Since 1839 very great improvements have been made in the harbour, a deep channel having been cut right up to the town, so that large vessels drawing 16 or 18 feet water ... are now able to discharge cargo at the new quays, which with splendid docks etc, have cost the corporation half a million of money.**

ABOVE: **Harland and Wolff's shipbuilding yard in Belfast.**

OVERLEAF: **Queen's University, Belfast.**

The contractor responsible for digging out that harbour silt was the ubiquitous 'King of the Railways' himself, William Dargan. The docks were now perfect for new shipbuilding ventures, and in 1858 Edward Harland, manager of a small yard on Queen's Island, bought out his employers and installed a colleague, Gustav Wolff, as his partner. Harland & Wolff went on to become arguably the biggest name in the industry (they built the White Star liners *Titanic*, *Olympic*, *Britannic* and *Oceanic*) and Harland knew from the start that the company's fortunes were inextricably linked to the railways. He lobbied for track to be laid straight to the quayside, giving his own yard and rivals such as McIlwaine & Coll easy access to imported heavy components while ensuring speedy docking for the big textile factories such as Irish Linen Mills, Blackstaff Falls Flax, Emersons, Greeves, New Northern and Springfield. In time Harland & Wolff branched out into designing railway locomotive engines using the labour skills and equipment it already used for shipbuilding.

Bradshaw is rather underwhelmed by Belfast city centre, remarking that:

> **The tall chimnies *[sic]* and factories for spinning linen and cotton yarn are the most conspicuous buildings; none of the churches are worth remark; in fact Belfast is a modern town, scarcely going back beyond the last century.**

While undeniably true, this dismissal of the city's architecture is not entirely fair. The architect Sir Charles Lanyon produced some outstanding work in the mid-nineteenth century. His great landmark structures include the main red-brick building at Queen's University (1849), the innovative Crumlin Road Gaol and Courthouse (1850), the imposing Italian Renaissance-style Customs House (1857) with its carved statues of Neptune, Mercury and Britannia, and The Abbey at Whiteabbey – a grand country residence built for a local MP but later acquired by Lanyon himself. But for the Victorian tourist, perhaps Lanyon's most celebrated achievement was the Palm House in Belfast's Botanic Gardens.

Established in 1828, these gardens became, and remain, a much-loved green oasis in the heart of the city. The Palm House, completed in 1852, was a particular favourite of the Victorians because its 'stove wing' utilised new hothouse technology to nurture some of the world's most exotic tropical plants. This elegant structure rises to a 46-ft-high elliptical dome and is one of the earliest examples of a curved iron glasshouse.

Early rail travellers to Belfast would have been familiar with the Railway Tavern in Great Victoria Street as a refreshment stop. However, this pub's glory days emerged only after it was taken over by Michael Flanagan in 1885, who renovated it as a fashionable 'gin palace' and renamed it the Crown Liquor Saloon. Gin palaces date from the mid-eighteenth century when gin, supposedly a medicine, was sold by chemists as a 'quick nip' on the spot or to take away. These gin-shop counters were designed for speedy service and later became a model for the traditional pub bar. By the 1820s they had become large and outrageously decorated licensed establishments, often illuminated by gaslight.

Though thought to be a vulgar haunt of the lower classes, gin palaces were massively popular; Charles Dickens described them as 'perfectly dazzling' in one of his literary sketches. In the case of the Crown Liquor Saloon, the wonderful tiling, woodwork and stained-glass windows are largely thanks to Flanagan's legendary charm – he talked Italian craftsmen engaged on the city's newly built churches into moonlighting after hours in his pub. Today customers sitting in the pub's discreet 'snug' bars can still see the bells used to summon drinks and the gunmetal plates conveniently sited for smokers to strike matches.

The extension of the Belfast & County Down railway to Downpatrick in 1859 should have allowed Bradshaw to wax lyrical on the town's ancient association with St Patrick. Ireland's foremost patron saint is reputed to be buried there, but Bradshaw notes only that Down Cathedral 'contains the tomb of Lord Kehany; the window at the east end is worth notice'. Whereas Kehany was a minor aristocrat, Patrick is reputed to have converted Ireland to Christianity in the fifth century. His hagiographies claim he could pass through locked doors, turn night into day and transform his followers into deer to protect them from enemies. Later legends tell how he banished all the snakes from Ireland.

Restoration of the original fourteenth-century Down Cathedral was completed in 1826, but the inscribed chunk of Mourne granite marking Patrick's grave wasn't placed in the grounds until 1900. Perhaps this is why Bradshaw overlooked the saint whose feast day on 17 March has since become a truly worldwide celebration.

While Downpatrick was a popular site with visitors it was the Antrim coast north of Belfast which saw an explosion in late Victorian rail tourism. This was great news for the Belfast & Northern Counties Railway and its chief engineer, Berkeley Deane Wise, in particular. Wise, who had cut his teeth on the railways as an assistant to Dargan and Brunel back on the Bray Head line, designed more than a dozen stations and hotels for his employers, the most famous of which is the mock-Tudor station at Portrush. This was built to help cope with unprecedented summer traffic to the north-west coast and still stands today.

RIGHT: At the Downpatrick Heritage Steam Railway.

OVERLEAF: Gobbins Path, the walk Wise constructed along the magnificent Gobbins cliffs.

Wise was also in the business of creating tourist attractions to drum up passenger numbers. His Gobbins Path at Whitehead, reached via the Carrickfergus & Larne rail line north of Belfast, comprised a remarkable series of tunnels, walkways and tubular bridges which stretched for two miles along spectacular cliffs. This pushed all the right buttons for the Victorians, combining a bracing seaside walk, wonderful sea views and breathtaking engineering. One advertisement held in the archives of Whiteabbey Presbyterian Church reads: 'New cliff path along the Gobbins Cliff, with its ravines, bore caves, natural aquariums etc, has no parallel in Europe as a marine cliff walk.'

Luring tourists to Antrim helped make the BNCR Ireland's most profitable railway, and Wise, backed by his general manager Edward Cotton, was given a free rein to pursue his ideas. As well as tea rooms, bandstands and golf courses, these included the Promenade in Whitehead, constructed using railway sleepers and set above a man-made beach, and the Blackhead Path (The Gobbins' 'sister-path') which stretched 1¼ miles out to Blackhead promontory. Among Wise's most celebrated tourist 'honey-pots' was a walk alongside the waterfalls at Glenariff, one of the nine Glens of Antrim, which included cantilevered cliff paths and picturesque natural shelters. Tourists ferried by carriage from the Parkmore narrow-gauge railway station could refresh themselves in a tea room below the Ess-na-Larach waterfall. This also offered budding landscape photographers a fully equipped darkroom.

The Gobbins was to prove almost as popular as the Giant's Causeway, near Coleraine to the north-west, another destination which boosted BNCR railway traffic out of Belfast. Sightseers heading to this dramatic volcanic rock formation near Bushmills would travel via Ballymena to Portrush Station near Coleraine before crossing Eglinton Street to board the Giant's Causeway, Portrush & Bush Valley Railway and Tramway. This was a revolutionary 3-ft-gauge electric railway partly powered by a hydro-electric turbine station at Walkmill Falls, Bushmills.

The line was fully open by July 1897 but suffered a setback eight years later when a cyclist was fatally electrocuted after touching the live conductor rail. A later inquiry established that voltage on the electrical feed varied between 290 and 360 volts and the company was compelled to agree a voltage reduction which restricted the number of services.

Dungiven Castle, formerly the ancestral home of the O'Cahan clan, Co. Derry.

An overhead conductor rail was installed, with mixed results, and it wasn't until 1907 that a reliable voltage of 550V was established.

Of the Causeway itself Bradshaw is suitably impressed, urging readers to head for the Causeway Inn and walk to Dunseverick Beach before taking a boat back, preferably with a guide (fee two shillings and sixpence). The advantage of this plan was that 'the succession of pillars and stratifications of the rocks along this remarkable coast are now fully visible'.

The final leg of the route goes west from Coleraine to Londonderry, or Derry as many prefer to call it. A few miles south of the railway lies Dungiven Castle from where the music to 'Londonderry Air', better known as 'Danny Boy', is said to originate. It is claimed the composer of this haunting melody (whose identity is lost in the mists of time) wrote it to mark the passing of the great Irish chieftain Cooeyna Gall O'Cahan, last of the O'Cahan chiefs, for whom Dungiven was the ancestral home. The music would have been played by Irish fiddlers in the nineteenth century but became internationally famous only after the Victorian era when the English lyricist Frederic Weatherly set it to the words of 'Danny Boy'.

And so to Londonderry, a city steeped in Georgian architecture after it was largely rebuilt during the eighteenth century in the wake of the 1689 Siege of Derry, another ignominious reversal for the ousted Catholic James II during the Glorious Revolution. The 105-day siege saw the deaths of some 8,000 inhabitants, almost a quarter of the total population, but the city was held by Protestant loyalists until it was relieved by a Royal Navy fleet. The quick thinking of 13 apprentice boys, who managed to lock the city gates against an advancing 1,200-strong Scottish Catholic army, is still celebrated today in a march by the Apprentice Boys of Derry.

Bradshaw describes Londonderry's 'considerable commercial intercourse with America and the West Indies, it being favourably situated for commerce, and possesses an excellent secure harbour with a splendid line of quays'. In fact, by the time his *Descriptive Railway Handbook* was published this 'commerce' was largely of the human variety; during the 1860s around 100,000 emigrants per year left Ireland for America, and from 1876 to 1921 this was the destination for 84 per cent of Irish emigrants, compared with 7 per cent heading for Canada and 8 per cent to the British mainland. As Evelyn Waugh later observed, there were only two final realities for the Irish: Hell and the United States. Famine was the driving force behind emigration, and although Londonderry was by no means the largest embarkation port its rail link made it a popular choice.

IT IS A BITTER IRONY THAT THE RAILWAY WHICH BROUGHT IRELAND SO MUCH WEALTH AND PROSPERITY ALSO DELIVERED ITS PEOPLE SO EFFICIENTLY INTO EXILE

The most famous of the city's shipping companies was the McCorkell Line and its services to New York, Philadelphia, New Orleans and Quebec were in huge demand during the famine years between 1845 and 1850. The *Mohongo*, a Canadian-built ship, completed more than 100 such crossings with no serious difficulty, while the *Minnehaha*, built in 1860, was able to cross the Atlantic quickly even during the winter months. She became the McCorkell Line's most celebrated vessel and passed into Irish-American folklore as the 'Green Yacht from Derry'.

For thousands of poverty-stricken emigrants the train ride to Londonderry would offer the last sights of home. It is a bitter irony that the railway which brought Ireland so much wealth and prosperity also delivered its people so efficiently into exile.

INDEX

Entries in *italics* indicates photographs and images

M

N

O

P

Q

R

S

T

U

V

W

Y

Z

PICTURE CREDITS

Endpapers: National Railway Museum/SSPL; 2 National Railway Museum/SSPL; 6-7, 8-9, 14-15 Old House books & maps; 16 © John Peter Photography/Alamy; 17 (top) © Science Museum Pictorial/Science & Society Picture Library; 17 (bottom) © Paul Mogford/Alamy; 19 © Amoret Tanner/Alamy; 21 The Stapleton Collection/The Bridgeman Art Library; 22 © Hulton-Deutsch Collection/CORBIS; 23 © Illustrated London News Ltd/Mary Evans; 24 © Look and Learn/The Bridgeman Art Library; 25 © Illustrated London News Ltd/Mary Evans; 26 © Illustrated London News Ltd/Mary Evans; 27 © Lordprice Collection/Alamy; 28–9 © Royal Mail Group Ltd/The Bridgeman Art Library; 32–3 Illustrated London News Ltd/Mary Evans; 35 Mary Evans Picture Library/Francis Frith; 36–7 © John Worrall/Alamy; 38 Library of Congress, Prints & Photographs Division; 40 Mary Evans Picture Library; 41 © Illustrated London News Ltd/Mary Evans; 42 © Alan Reed/Alamy; 43 © Tom Mackie/Alamy; 44–5 Library of Congress, Prints & Photographs Division; 46–7 © Illustrated London News Ltd/Mary Evans; 49 © Paul Heinrich/Alamy; 51 © 19th era/Alamy; 52–3 Library of Congress, Prints & Photographs Division; 55 © Illustrated London News Ltd/Mary Evans; 57 Mary Evans Picture Library; 58 © Stephen Sykes/Alamy; 60 Mary Evans Picture Library; 61 Getty Images; 62 Getty Images; 63 Mary Evans Picture Library; 64 © Paul Tavener/Alamy; 65 Mary Evans/Peter Higginbotham Collection; 66–7 National Railway Museum/SSPL; 70–1 National Railway Museum/SSPL; National Railway Museum/The Bridgeman Art Library; 74 © Tarker/The Bridgeman Art Library; 77 © Illustrated London News Ltd/Mary Evans; 78–9 Mary Evans Picture Library; 118–19 Mary Evans Picture Library; 81 Library of Congress, Prints & Photographs Division; 82 © Lordprice Collection/Alamy; 83 © Illustrated London News Ltd/Mary Evans; 84–5 © Bernie Epstein/Alamy; 86 © The Print Collector/Alamy; 87 Press Association Images; 89 © Bettmann/CORBIS; 91 © The Print Collector/Alamy; 92–3 Library of Congress, Prints & Photographs Division; 95 Mary Evans Picture Library; 96 © Les Ladbury/Alamy; 97 Mary Evans Picture Library; 99 Mary Evans/Roger Worsley Archive; 101 Library of Congress, Prints & Photographs Division; 102 © CORBIS; 107 © The Photolibrary Wales/Alamy; 108–9 Library of Congress, Prints & Photographs Division; 111 Mary Evans Picture Library; 112 © Lordprice Collection/Alamy; 113 © Mary Evans Picture Library/Alamy; 115 Mary Evans Picture Library; 116 © Archive Images/Alamy; 117 © Lebrecht Music and Arts Photo Library/Alamy; 121 © Illustrated London News Ltd/Mary Evans; 122–3 Mary Evans Picture Library; 124 Guildhall Library, City of London/The Bridgeman Art Library; 125 © Greg Balfour Evans/Alamy; 127 © Illustrated London News Ltd/Mary Evans; 128 Mary Evans Picture Library; 129 © Mary Evans Picture Library/Alamy; 130 © Michael Rose/Alamy; 131 © Bettmann/CORBIS; 132–3 © Johnny Jones/Alamy; 134 National Railway Museum/SSPL; 137 Mary Evans Picture Library; 139 © Motoring Picture Library/Alamy; 140–1 © Peter Titmuss/Alamy; 145 © Illustrated London News Ltd/Mary Evans; 146 © Patrick Eden/Alamy; 148–9 © Roy Rainford/Robert Harding World Imagery/Corbis; 150–1 Library of Congress, Prints & Photographs Division; 153 © Peter Titmuss/Alamy; 157 © Susanne Masters/Alamy; 158–9 © Steve Taylor ARPS/Alamy; 160–1 Mary Evans Picture Library; 163 © Illustrated London News Ltd/Mary Evans; 164–5 © Roger Bamber/Alamy; 166 Mary Evans Picture Library/Francis Frith; 167 © Pictorial Press Ltd/Alamy; 170 Mary Evans Picture Library; 171 © World History Archive/Alamy; 173 © World History Archive/Alamy; 174 Mary Evans Picture Library/Onslow Auctions Limited; 175 Getty Images; 177 © Mary Evans Picture Library/Alamy; 178 © Pictorial Press Ltd/Alamy; 179 © World History Archive/Alamy; 180 National Railway Museum/SSPL; 183 National Railway Museum/SSPL; 184 Mary Evans/The National Archives, London. England; 186–7 © The Print Collector/Alamy; 188 Mary Evans/Grenville Collins Postcard Collection; 191 © David Taylor Photography/Alamy; 192 © World History Archive/Alamy; 194–5 © Jon Sparks/Alamy; 196 © Patrick Nairne/Alamy; 197 © Bettmann/CORBIS; 198 ©Thomas Cook Archive/Illustrated London News Ltd/Mary Evans; 201 Mary Evans Picture Library; 202 © Colin Palmer Photography/Alamy; 205 © International Photobank/Alamy; 206–7 © 19th era/Alamy; 208 © B. O'Kane/Alamy; 210–11 © JoeFox CountyDown/Alamy; 212–13 Private Collection/The Stapleton Collection/The Bridgeman Art Library; 214 Mary Evans Picture Library; 217 National Railway Museum/SSPL; 218 © Design Pics Inc. – RM Content/Alamy; 221 National Railway Museum/SSPL; 222–3 © Liszt Collection/Alamy; 224–5 © Isifa Image Service S.R.O./Alamy; 226 © Lonely Planet Images/Alamy; 230–1 Library of Congress, Prints & Photographs Division; 233 © Barry Mason/Alamy;235 Mary Evans Picture Library; 236 © 19th Era 2/Alamy; 239 Mary Evans Picture Library; 241 Getty Images; 243 © Mary Evans Picture Library/Alamy; 244–5 © Macduff Everton/Corbis; 248–9 © Sebastian Wasek/Alamy; 250 © David Taylor Photograph/Alamy

Robert Stephenson Specification 22 December 1841 Improvements in the arrangement and combination of the parts of
Lateral Elevation
Figure 1
Safety Valve
Fire box
A
Cylindrical part of the boiler
C
spring
frame
U
pump rod 16
k axis
excentric rod 3
connecting rod
suction pipe
m axis
Hindmost wheel
M
Main wheel
K
Robt Stephenson
Iron rails
These dotted lines shew the place of the hindmost wheel
These dotted lines shew the place of the Main wheel
spring
crank
Connecting rod
Dome or steam head
Z
pump
pump rod
half
hoop
excentric rods
A
C
over the fire box
Figure 2 Horizontal Plan
Scale of Feet and Inches
Inches